SEQUENCES OF REAL AND COMPLI

$$e = \lim_{n \to +\infty} \left(1 + \frac{1}{n}\right)^n$$

$$Golden\ Ratio\ \phi = \frac{\sqrt{5} + 1}{2}$$

$$= \sqrt{1 + \sqrt{1 + \sqrt{1 + \sqrt{1 + \cdots}}}}$$

1) A concise and complete introduction to the Theory of Infinite Sequences of Real and Complex Numbers.

2) An excellent supplementary text for all Mathematics, Physics and Engineering students.

3) 93 solved illustrative examples and 260 characteristic problems to be solved.

4) Odd numbered problems are provided with answers.

About the Author

Demetrios P. Kanoussis, Ph.D

Kalamos Attikis, Greece

dkanoussis@gmail.com

Dr. Kanoussis is a professional Electrical Engineer and Mathematician. He received his Ph.D degree in Engineering and his Master degree in Mathematics from Tennessee Technological University, U.S.A, and his Bachelor degree in Electrical Engineering from the National Technical University of Athens (N.T.U.A), Greece.

As a professional Electrical Engineer, Dr. Kanoussis has been actively involved in the design and in the implementation of various projects, mainly in the area of the **Integrated Control Systems.**

Regarding his teaching experience, Dr. Kanoussis has long teaching experience in the field **of Applied Mathematics and Electrical Engineering.**

His original scientific research and contribution, in Mathematics and Electrical Engineering, is published in various, high impact international journals.

Additionally to his professional activities, teaching and research, Demetrios P. Kanoussis is **the author of several textbooks in Electrical Engineering and Applied Mathematics.**

A complete list of Dr. Kanoussis textbooks in Mathematics and Engineering can be found in the Author's page at Amazon Author

Central (https://www.amazon.com/Demetrios-P.-Kanoussis/e/B071GZ215Z)

Sequences of Real and Complex Numbers.

Copyright 2017,   **Author**: Demetrios P. Kanoussis.

First edition, March 2017.

Preface

This book is a complete and self contained presentation of the fundamentals of Sequences of real and complex numbers, and is intended primarily for students of Sciences and Engineering. Infinite Sequences Theory is an important tool for all Science and Engineering students.

Sequences, in a sense, constitute an introduction to the so called "**Higher Mathematics**". The notion of the **limit**, which is a core, fundamental concept in the study of many areas of Advanced Mathematics, Physical Sciences and Engineering, is introduced in sequences.

Many important areas in Mathematics, with a wide range of applications, in Physical Sciences and Engineering, like Infinite Series, Derivatives ,Integrals, etc, rely heavily on the notion of the limit, and therefore on sequences.

This textbook is written to provide any possible assistance to the students who are first being introduced to the theory of sequences, but it could, equally well, be used by students already exposed to the theory and wishing to broaden their theoretical background and analytical skills on the subject.

The content of this book is divided into 16 chapters, as shown in the table of contents.

The 93 solved illustrative examples and the 260 characteristic problems to be solved, are designed to help students gain confidence and enhance their understanding, working knowledge and computational skills on sequences.

A brief **hint** or a **detailed outline** of the procedure to be followed, in solving more complicated problems, is often given.

Finally, **answers** to the odd numbered problems, are also given, so that the students can easily verify the validity of their own solution.

Demetrios P. Kanoussis

Table of contents

1. Basic Concepts and Definitions.

Let $\mathbb{N}$ be the set of natural numbers ($\mathbb{N}= \{1, 2, 3, ...\}$) and $\mathbb{R}$ be the set of real numbers.

A sequence (or infinite sequence) of real numbers, is a function having domain $\mathbb{N}$ and range $\mathbb{R}$. **In other words, a sequence is a rule that assigns to each element in the set $\mathbb{N}$, one and only one element in the set $\mathbb{R}$.**

$$
\begin{array}{ccccccc}
1 & 2 & 3 & \cdots & n & n+1 & \cdots \\
\downarrow & \downarrow & \downarrow & \downarrow & \downarrow & \downarrow & \downarrow \\
x_1 & x_2 & x_3 & \cdots & x_n & x_{n+1} & \cdots
\end{array}
$$

Figure 1-1: Sequence of real numbers.

The numbers x_1, x_2, x_3, ..., x_n, ... are the **terms** or the **elements** of the sequence, x_1 being the first term, x_2 the second term, x_3 the third term, ... , and in general , x_n the nth term or the general term of the sequence.

If we call f the rule, which assigns to each element n in $\mathbb{N}$ one and only one element x_n in $\mathbb{R}$, we may write:

$$n \in \mathbb{N} \xrightarrow{f} x_n = f(n) \in \mathbb{R} . \tag{1-1}$$

We may express the sequence in (1-1), as $\{x_1, x_2, x_3, x_4, \cdots, x_n, x_{n+1} \cdots\}$ or simply as (x_n).

The simplest way to know the terms of a sequence is to know **how the nth term x_n is expressed, in terms of n, in other words to know the function $x_n = f(n)$, where $f(n)$ is a known function of n,** for instance, $x_n = n + 5$, or $x_n = \sqrt[3]{n^2 + 1}$, etc.

Another way to define a sequence, **is to know the first term x_1, and a mathematical relationship, by means of which each term of the sequence (for $n \geq 2$) is defined in terms of its predecessor.** In this case,

$x_{n+1} = f(x_n)$, x_1 is given, $\qquad\qquad\qquad\qquad$ (1-2)

and we say that **the sequence (x_n) is defined recursively.**

Equation (1-2) defines **a first order recursive sequence.**

In a quite similar manner, one may define a second order recursive sequence,

$x_{n+2} = f(x_{n+1}, x_n)$, **where those x_1, x_2 are given.** $\qquad$ (1-3)

Equation (1-3) defines **a second order recursive sequence.** In this case, the terms are

$$x_3 = f(x_2, x_1), \quad x_4 = f(x_3, x_2), \quad x_5 = f(x_4, x_3), \cdots \text{ e.t.c.}$$

If a sequence is defined by means of a recursive formula, it is possible, in some cases, to be able to express the $n^{\underline{th}}$ term x_n, in terms of n. From a theoretical and practical point of view, this is desirable (see Example 1-4).

Recursive sequences are studied in detail, in Chapter 12.

Note: The Σ and Π notation.

Let $x_1, x_2, x_3, \cdots, x_n$ be n numbers (real or complex).

The sum $x_1 + x_2 + x_3 + \cdots + x_n$, can be expressed briefly using the Greek symbol Σ as $\sum_{k=1}^{n} x_k = x_1 + x_2 + x_3 + \cdots + x_n$. $\qquad\qquad$ (1-4)

The index k is a dummy variable, in the sense that

$$\sum_{k=1}^{n} x_k = \sum_{m=1}^{n} x_m = \sum_{i=1}^{n} x_i = x_1 + x_2 + x_3 + \cdots + x_n.$$

The sum in (1-4) could also be written, as

$$\sum_{k=2}^{n+1} x_{k-1} \ or \ \sum_{m=3}^{n+2} x_{m-2} \ or \ \sum_{i=4}^{n+3} x_{i-3} = x_1 + x_2 + x_3 + \cdots + x_n.$$

Similarly, the product $x_1 \cdot x_2 \cdot x_3 \cdots x_n$, can be expressed briefly using the Greek symbol Π as $\prod_{k=1}^{n} x_k = x_1 \cdot x_2 \cdot x_3 \cdots x_n$. (1-5)

Obviously, k is a dummy variable, again, since

$$\prod_{k=1}^{n} x_k = \prod_{m=1}^{n} x_m = \prod_{i=1}^{n} x_i = x_1 \cdot x_2 \cdot x_3 \cdots x_n.$$

The product in (1-5) could also be written as

$$\prod_{k=2}^{n+1} x_{k-1} \; or \; \prod_{m=3}^{n+2} x_{m-2} \; or \; \prod_{i=4}^{n+3} x_{i-3} = x_1 \cdot x_2 \cdot x_3 \cdots x_n.$$

It is easy to show that if c and d are any two constants, then

$$\sum_{k=1}^{n}(c \cdot x_k + d \cdot y_k) = c \cdot \sum_{k=1}^{n} x_k + d \cdot \sum_{k=1}^{n} y_k .$$ (1-6)

Example 1-1.

Write the first five terms of each one of the following sequences:

a) $x_n = \frac{n-1}{n}$,   **b)** $y_n = \frac{\cos(n\pi)}{n}$,   **c)** $w_n = \left(1 + \frac{1}{n}\right)^n$.

Solution

a) $x_1 = 0$, $x_2 = \frac{1}{2}$, $x_3 = \frac{2}{3}$, $x_4 = \frac{3}{4}$, $x_5 = \frac{4}{5}$.

b) $y_1 = -1$, $y_2 = \frac{1}{2}$, $y_3 = -\frac{1}{3}$, $y_4 = \frac{1}{4}$, $y_5 = -\frac{1}{5}$.

c) $w_1 = 2$, $w_2 = \left(\frac{3}{2}\right)^2$, $w_3 = \left(\frac{4}{3}\right)^3$, $w_4 = \left(\frac{5}{4}\right)^4$, $w_5 = \left(\frac{6}{5}\right)^5$.

Example 1-2.

Write the first five terms of the first order recursive sequence, $x_{n+1} = x_n + \frac{1}{x_n}$, $x_1 = 1$.

Solution

$x_1 = 1$, (given),

$$x_2 = x_1 + \frac{1}{x_1} = 1 + 1 = 2,$$

$$x_3 = x_2 + \frac{1}{x_2} = 2 + \frac{1}{2} = \frac{5}{2} \ ,$$

$$x_4 = x_3 + \frac{1}{x_3} = \frac{5}{2} + +\frac{2}{5} = \frac{29}{10} \ ,$$

$$x_5 = x_4 + \frac{1}{x_4} = \frac{29}{10} + \frac{10}{29} = \frac{941}{290} \ .$$

Example 1-3.

Write the first five terms of the second order recursive sequence,

$$x_{n+2} = x_{n+1} + x_n , \qquad x_1 = 1 , x_2 = 1.$$

Solution

$x_1 = 1, \ x_2 = 1$, (given),

$x_3 = x_2 + x_1 = 1 + 1 = 2$,

$x_4 = x_3 + x_2 = 2 + 1 = 3$,

$x_5 = x_4 + x_3 = 3 + 2 = 5.$

The given recursive sequence is known as the **Fibonacci sequence**, since it was introduced firstly by **Fibonacci (1175-1250)**, also known as **Leonardo of Pisa**, while investigating a problem involving the offspring of rabbits. There exists an extensive mathematical literature, about **the Fibonacci numbers**, i.e. the numbers generated by the Fibonacci sequence.

Example 1-4.

Consider the sequence $\{x_{n+1} = 2 \cdot \frac{1+x_n}{3+x_n} \ , \ x_1 = 3 \}$ and determine the term

x_{1000} .

Solution

The given sequence is **a first order recursive sequence.**

In order to determine the term x_{1000} , one has to determine **all the in between terms, x_2 , x_3 , x_4 , x_5 , ..., x_{998} , x_{999}** , which obviously would be quite complicated and very time consuming, (almost impossible!).

However, as it will be shown, in the sequel, (see Example 12-3), **one may express the term x_n in terms of n**, the result being,

$$x_n = \frac{5 \cdot 4^{n-1} + 4}{5 \cdot 4^{n-1} - 2} , \qquad n = 1,2,3,...$$

We notice that for $n = 1$, we recover the first term $x_1 = 3$. The term x_{1000} will be,

$$x_{1000} = \frac{5 \cdot 4^{999} + 4}{5 \cdot 4^{999} - 2} .$$

Example 1-5.

If c is any constant, $(c \neq 0)$, show that $\prod_{i=1}^{n}(c \cdot x_i) = c^n \cdot \prod_{i=1}^{n} x_i$.

Solution

$$\prod_{i=1}^{n}(c \cdot x_i) = (cx_1)(cx_2)(cx_3) \cdots (cx_n) = c^n x_1 x_2 x_3 \cdots x_n = c^n \prod_{i=1}^{n} x_i.$$

PROBLEMS

1-1) Write the first five terms of the sequence $x_n = \frac{n}{2^n}$.

(Answer: $x_1 = \frac{1}{2}$, $x_2 = \frac{2}{4}$, $x_3 = \frac{3}{8}$, $x_4 = \frac{4}{16}$, $x_5 = \frac{5}{32}$).

1-2) If $x_n = \dfrac{(1+\sqrt{5})^n - (1-\sqrt{5})^n}{\sqrt{5}\cdot 2^n}$, $n = 1,2,3,\dots$, show that $x_1 = 1$, $x_2 = 1$, and that $x_{n+2} = x_{n+1} + x_n$. Relate this problem with Example 1-3, and draw the appropriate conclusions.

1-3) Write the first five terms of the sequence $\{\, x_{n+1} = \sqrt{1 + \dfrac{1}{(x_n)^2}}\ ,\ x_1 = 1 \,\}$.

(Answer: $x_1 = 1$, $x_2 = \sqrt{2}$, $x_3 = \sqrt{\dfrac{3}{2}}$, $x_4 = \sqrt{\dfrac{5}{3}}$, $x_5 = \sqrt{\dfrac{8}{5}}$).

1-4) If $x_n = \dfrac{\cos(n\pi)}{n^2+1}$, find the terms x_3, x_7, x_8, x_{10}.

1-5) In Example 1-4, find the first five terms,

a) Using the recursive formula, and

b) Using the formula obtained, expressing x_n in terms of n.

(Answer: $x_1 = 3$, $x_2 = \dfrac{4}{3}$, $x_3 = \dfrac{14}{13}$, $x_4 = \dfrac{54}{53}$, $x_5 = \dfrac{214}{213}$).

1-6) Consider the sequence (x_n), defined as $\left\{ x_{n+2} = \dfrac{x_{n+1}+x_n}{2} , x_1 = c , x_2 = d \right\}$, where c and d are given. Show that, $x_{n+2} - x_{n+1} = \left(-\dfrac{1}{2}\right)^n \cdot (d-c)$.

1-7) If $\left\{ x_{n+1} = \dfrac{x_n}{1+x_n} ,\ x_1 \text{ given} \right\}$, show that

$$\sum_{k=1}^{n+1} \frac{1}{(x_k)^2} = \frac{1}{(x_1)^2} + \frac{1}{(x_2)^2} + \cdots + \frac{1}{(x_{n+1})^2} = \frac{n+1}{(x_1)^2} + \frac{n\cdot(n+1)}{x_1} + \frac{n\cdot(n+1)\cdot(2n+1)}{6}.$$

Hint: In order to work Problem 1-7, one may use the following identities:

a) $\sum_{k=1}^{n} k = 1 + 2 + 3 + \cdots + n = \dfrac{n(n+1)}{2}$, and

b) $\sum_{k=1}^{n} k^2 = 1^2 + 2^2 + 3^2 + \dots + n^2 = \dfrac{n(n+1)(2n+1)}{6}$.

Also, note that $\dfrac{1}{x_{n+1}} - \dfrac{1}{x_n} = 1$, $n = 1,2,3,\cdots$

1-8) The sequence (y_n) is defined by the following rule: y_n is the $n^{\underline{th}}$ digit of the number $\pi = 3.14159265 \cdots$. Write the first six terms of (y_n).

1-9) Find the general term x_n of the sequences,

a) $-\dfrac{1}{5}, \dfrac{3}{8}, -\dfrac{5}{11}, \dfrac{7}{14}, -\dfrac{9}{17}, \cdots$

b) $\dfrac{1}{2}, 1, \dfrac{9}{8}, 1, \dfrac{25}{32}, \dfrac{36}{64}, \cdots$

(**Answer: a)** $(-1)^n \dfrac{2n+1}{3n+2}$ **b)** $\dfrac{n^2}{2^n}$)

1-10) Consider the sequence (x_n), with general term $x_n = n^3 - 2n^2 - n + 2$. For which value of n, the corresponding term of the sequence equals 8?

(**Answer:** $n = 3$).

1-11) Consider the sequence $\{ x_{n+1} = \dfrac{1}{2}\left(x_n + \dfrac{9}{x_n}\right) , x_1 = 3 \}$, and show that $x_1 = x_2 = x_3 = \cdots = 3$, i.e. $x_n = 3$, $\forall n \in \mathbb{N}$, $((x_n)$ **is a constant sequence**).

1-12) Write down the first six terms of each of the sequences defined below:

a) $x_n = \left(\sin\left(\dfrac{n\pi}{2}\right)\right)^n$,   **b)** $y_n = \dfrac{n^3+1}{n^2+2}$,   **c)** $a_n = \sqrt{n+1} - \sqrt{n}$,

d) $\left\{ b_{n+1} = b_n + \left(\dfrac{1}{3}\right)^n , \ b_1 = 1 \right\}$,   **e)** $\left\{ c_{n+1} = \dfrac{c_1+c_2+\cdots+c_n}{n+1} , \ c_1 = 1 \right\}$.

1-13) If $x_n = \left(\dfrac{1}{3}\right)^n$, how large n must be, for $x_n < 10^{-4}$?

(**Answer:** $n \geq 9$).

1-14) If $y_n = 1.2^n$, how large n must be, for $y_n > 350$?

1-15) If $x_n = 2n + 1$, find the value of n, for which $\left(\dfrac{x_{n+1}}{x_n}\right)^2 = \dfrac{81}{49}$?

(**Answer:** $n = 3$).

1-16) Show that $\prod_{n=1}^{k} \cos(2^n x) = \dfrac{1}{2^k} \cdot \dfrac{\sin(2^{k+1}x)}{\sin(2x)}$.

Hint: Make use of the Trigonometric identity, $\cos x = \dfrac{\sin 2x}{2 \sin x}$

2. Bounded Sequences.

Definition 2-1: A real sequence (x_n), is said to be **bounded from the right, or bounded above**, if a constant number k exists, such that for all $\in \mathbb{N}$,

$$x_n \leq k, \quad \forall n \in \mathbb{N}, \qquad (2\text{-}1)$$

The number k is also known as **an upper bound of (x_n).** The upper bound k (if it exists), is not unique. Obviously, any other number $k_1 > k$ may also be considered as an upper bound of (x_n).

Definition 2-2: A real sequence (x_n), is said to be **bounded from the left, or bounded below**, if a constant number ℓ exists, such that for all $n \in \mathbb{N}$,

$$x_n \geq \ell, \quad \forall n \in \mathbb{N}. \qquad (2\text{-}2)$$

The number ℓ is also known as **a lower bound of (x_n).** The lower bound ℓ (if it exists), is not unique, since any other number $\ell_1 < l$, will also be a lower bound of (x_n).

Definition 2-3: A real sequence (x_n), is said to be **absolutely bounded**, if a positive number $m > 0$ exists, such that for all $n \in \mathbb{N}$,

$$|x_n| \leq m, \quad \forall n \in \mathbb{N}. \qquad (2\text{-}3)$$

Theorem 2-1.

Every absolutely bounded sequence (x_n) has an upper and a lower bound, and conversely, every sequence (x_n) having an upper and a lower bound, is absolutely bounded.

Proof: Assuming that (x_n) is absolutely bounded, $|x_n| \leq m$, $\forall n \in \mathbb{N}$, or equivalently,

$-m \leq x_n \leq m$, $\forall n \in \mathbb{N}$, showing that (x_n) has the positive number m as an upper bound, and the negative number $(-m)$ as a lower bound.

Conversely, let's assume that k and ℓ are the upper and lower bound, respectively, of a real sequence (x_n), meaning that,

$$\ell \leq x_n \leq k, \ \forall n \in \mathbb{N}. \tag{2-4}$$

If $= max\{|k|, |\ell|\}$, then (2-4) implies that $-M \leq x_n \leq M$, $\forall n \in \mathbb{N}$,

meaning that $|x_n| \leq M$, $\forall n \in \mathbb{N}$, therefore (x_n) is absolutely bounded, and the proof is completed.

Note: A sequence (x_n) having upper and lower bound, is called **a bounded sequence**. By virtue of Theorem 2-1, every absolutely bounded sequence is a bounded sequence, and conversely, every bounded sequence is also, absolutely bounded.

Example 2-1.

Show that the sequence $x_n = \frac{1}{n}$, $n = 1,2,3, ...$, is bounded.

Solution

The first few terms of (x_n) are, $x_1 = 1, x_2 = \frac{1}{2}, x_3 = \frac{1}{3}, \cdots$

Obviously, all terms of (x_n) are positive and less than 1, i.e.

$$0 \leq x_n \leq 1, \ \forall n \in \mathbb{N}. \quad (*)$$

Therefore, the number 0 is a lower bound of (x_n) and the number 1 is an upper bound of (x_n), i.e. the sequence (x_n) is bounded.

Note: Any negative number is also a lower bound of (x_n), while any positive number greater than 1 is also an upper bound of (x_n). **The lower and the upper bound of a bounded sequence (x_n), are not unique.**

Also, from (*) one easily obtains that $-1 \leq x_n \leq 1$, or equivalently, $|x_n| \leq 1$, $\forall n \in \mathbb{N}$, confirming thus Theorem 2-1.

Example 2-2.

Show that the sequence $x_n = \frac{(-1)^n}{n}$, $n = 1,2,3, \cdots$ is bounded.

Solution

The first few terms of the given sequence are $x_1 = -1$, $x_2 = \frac{1}{2}$, $x_3 = -\frac{1}{3}$,

$x_4 = \frac{1}{4}$, $x_5 = -\frac{1}{5}, \cdots$

The sequence (x_n) is alternating. Obviously,

$-1 \leq x_n \leq \frac{1}{2}$, $\forall n \in \mathbb{N}$,

meaning that (x_n) is bounded, (and therefore absolutely bounded, for example $|x_n| \leq 1, \forall n \in \mathbb{N}$).

Example 2-3.

Show that the sequence $x_n = \frac{sin(n^3 + 5n^2 + 10)}{n^4 + 7}$ is bounded.

Solution

$|x_n| = \frac{|sin(n^3 + 5n^2 + 10)|}{|n^4 + 7|} \leq \frac{1}{n^4 + 7} < 1$, $\forall n \in \mathbb{N}$,

meaning that the sequence (x_n) is absolutely bounded, and therefore bounded, (by virtue of Theorem 2-1).

Example 2-4.

Show that the sequence $x_n = \frac{1}{n+1} + \frac{1}{n+2} + \frac{1}{n+3} + \cdots + \frac{1}{2n}$ is bounded.

Solution

Obviously $x_n > 0$, $\forall n \in \mathbb{N}$, i.e. the number 0 is a lower bound of (x_n). Also,

$x_n = \frac{1}{n+1} + \frac{1}{n+2} + \frac{1}{n+3} + \cdots + \frac{1}{2n} < \frac{1}{n+1} + \frac{1}{n+1} + \frac{1}{n+1} + \cdots \frac{1}{n+1} = \frac{n}{n+1} < 1$,

showing that the number 1 is an upper bound of (x_n).

Finally, $0 < x_n < 1$, $\forall n \in \mathbb{N}$, and this shows that (x_n) is bounded.

Example 2-5.

If $c > 0$, and $\{x_{n+1} = \sqrt{x_n + c}, \quad x_1 = \sqrt{c}\}$, show that $(c + 1)$ is an upper bound for the sequence (x_n).

Solution

We shall use **Mathematical Induction**, to work out Example 2-5.

We want to show that $x_n < c + 1$, $\forall n \in \mathbb{N}$.

a) For $n = 1$, $x_1 < c + 1 \Leftrightarrow \sqrt{c} < c + 1 \Leftrightarrow c < c^2 + 2c + 1 \Leftrightarrow 0 < c^2 + c + 1$, which is obviously true, since by assumption $c > 0$, and $c^2 + 1 > 1$.

b) We next assume that for some positive integer $k, x_k < c + 1$.

c) Finally, we shall show that $x_{k+1} < c + 1$, as well. Indeed,

$x_{k+1} < c + 1 \Leftrightarrow \sqrt{x_k + c} < c + 1 \Leftrightarrow x_k + c < c^2 + 2c + 1 \Leftrightarrow x_k < c^2 + c + 1.$

However, the last inequality is true, since by assumption (b), $x_k < c + 1$, while c^2 is a positive number.

According to **the principle of Mathematical Induction** $x_n < c + 1$, $\forall n \in \mathbb{N}$ and the proof is completed.

PROBLEMS

2-1) Show that the sequence $x_n = \dfrac{\cos(n^2+5)\cdot\sin(n^2+7)}{n^3+5}$ is bounded.

2-2) Show that the sequence $x_n = \dfrac{\cos(n^2+1)}{n^2+1} + \dfrac{\sin(n^3+2)}{n^3+2}$ is absolutely bounded.

2-3) If c is a constant > 1, and $\{x_{n+1} = \sqrt{c \cdot x_n}, \ x_1 = 1\}$, show that $x_n < c$, $\forall n \in \mathbb{N}$.

2-4) If d is a constant > 1, and $c = d^3 - d$, show that d is an upper bound of the sequence $\{x_{n+1} = \sqrt[3]{c + x_n}, x_1 = \sqrt[3]{c}\}$.

2-5) If $x_n = \frac{3n+5}{7n+4}$ show that the sequence (x_n) is bounded.

2-6) Consider the sequence with general term $x_n = \frac{1}{1^2} + \frac{1}{2^2} + \cdots \frac{1}{n^2}$ and show that (x_n) is bounded.

Hint: $\frac{1}{n^2} < \frac{1}{n-1} - \frac{1}{n}$, $n = 2,3,4, \cdots$

2-7) Consider the sequence $x_n = \frac{1}{1^3} + \frac{1}{2^3} + \frac{1}{3^3} + \cdots + \frac{1}{n^3}$, and show that (x_n) is bounded.

Hint: $\frac{1}{n^3} < \frac{1}{n^2}$, $n = 2,3,4, \cdots$

2-8) If $x_n = \left(1 - \frac{1}{2}\right) \cdot \left(1 - \frac{1}{3}\right) \cdot \left(1 - \frac{1}{4}\right) \cdots \left(1 - \frac{1}{n}\right)$, $n = 2,3,4, \cdots$ show that (x_n) is bounded.

Hint: $x_n = \frac{2-1}{2} \cdot \frac{3-1}{3} \cdot \frac{4-1}{4} \cdots \frac{n-1}{n} = \frac{1}{n}$

2-9) Show that the sequence $y_n = \frac{(\cos n)^2}{n^2} + \frac{\sin(n^3)}{n^3}$ is absolutely bounded.

2-10) Show that the sequence $w_n = \frac{7(\cos(n+1))^2 + 3(\sin(2n+5))^3}{n}$ is absolutely bounded.

2-11) Prove that the sequence $y_n = \frac{1+(-1)^n}{n^3}$ is bounded.

2-12) Which of the following sequences are bounded?

a) $x_n = \cos\frac{3n+1}{7n^2+5}$,   **b)** $y_n = 2^n + \frac{1}{n}$.

2-13) Prove that the sequence $x_n = \frac{7n}{2n+3}$ is bounded.

2-14) Prove that the sequence $y_n = \frac{3^n+1}{3^n}$ is bounded.

2-15) If (x_n) and (y_n) are two bounded sequences, show that the sequence $(x_n y_n)$ is also bounded.

3. Monotone Sequences.

Definition 3-1: A sequence (x_n), is called
Increasing if $x_{n+1} > x_n$, $\forall n \in \mathbb{N}$, i.e. if $\quad x_1 < x_2 < x_3 < \cdots < x_n < \cdots$, and
Decreasing if $x_{n+1} < x_n$, $\forall n \in \mathbb{N}$, i.e. if $x_1 > x_2 > x_3 > \cdots > x_n > \cdots$.
If $x_{n+1} \geq x_n$, $\forall n \in \mathbb{N}$, the sequence (x_n) is called **non-decreasing**, while if
$x_{n+1} \leq x_n$, $\forall n \in \mathbb{N}$, the sequence (x_n) is called **non-increasing**.

Definition 3-2: A sequence which is either increasing or decreasing, is called **a strictly monotone sequence**, while a sequence which is either non-decreasing or non-increasing is called simply **a monotone sequence**.

A usual method **to investigate a sequence with regard to its monotonicity**, i.e. to determine whether (x_n) is either increasing or decreasing, is to consider the difference $(x_{n+1} - x_n)$.
If $x_{n+1} - x_n > 0$, $\forall n \in \mathbb{N} \Leftrightarrow x_{n+1} > x_n$, $\forall n \in \mathbb{N}$, $\qquad$ (3-1)

the sequence (x_n) is increasing, and we write $(x_n) \nearrow$.

If $x_{n+1} - x_n < 0$, $\forall n \in \mathbb{N} \Leftrightarrow x_{n+1} < x_n$, $\forall n \in \mathbb{N}$, $\qquad$ (3-2)
the sequence (x_n) is decreasing, and we write $(x_n) \searrow$.

In the particular case where **all the terms of (x_n) are positive**, we may consider the ratio $(\frac{x_{n+1}}{x_n})$.
If $\frac{x_{n+1}}{x_n} > 1$, $\forall n \in \mathbb{N} \Leftrightarrow x_{n+1} > x_n$, $\forall n \in \mathbb{N} \Leftrightarrow (x_n) \nearrow$. (3-3)
If $\frac{x_{n+1}}{x_n} < 1$, $\forall n \in \mathbb{N} \Leftrightarrow x_{n+1} < x_n$, $\forall n \in \mathbb{N} \Leftrightarrow (x_n) \searrow$. (3-4)

The following two **Bernoulli inequalities** are often used, in order to show that a given sequence (x_n) is increasing or decreasing, (for a proof see Problem 3-16).

$(1 + a)^n > 1 + n \cdot a$, **where $a > 0$ and $n = 2, 3, 4, \cdots$.** $\qquad$ (3-5)
$(1 - a)^n > 1 - n \cdot a$, **where $0 < a < 1$ and $n = 2, 3, 4, \cdots$.** (3-6)

Example 3-1.

Show that the sequence $x_n = \frac{1}{n}$, $n \in \mathbb{N}$ is decreasing.

Solution

All the terms of (x_n) are positive numbers. The ratio

$$\frac{x_{n+1}}{x_n} = \frac{\frac{1}{n+1}}{\frac{1}{n}} = \frac{n}{n+1} < 1, \quad \forall n \in \mathbb{N},$$

Therefore $x_{n+1} < x_n$, $\forall n \in \mathbb{N}$, and the sequence (x_n) is decreasing $((x_n) \searrow)$.

Example 3-2.

Show that the sequence $x_n = \left(1 + \frac{1}{n}\right)^n$, $n \in \mathbb{N}$, is increasing.

Solution

All the terms of (x_n) are positive numbers. In order to show that the sequence is increasing, it suffices to show that,

$$\frac{x_{n+1}}{x_n} > 1 \Leftrightarrow \frac{\left(1+\frac{1}{n+1}\right)^{n+1}}{\left(1+\frac{1}{n}\right)^n} > 1 \Leftrightarrow \frac{\left(1+\frac{1}{n+1}\right)^{n+1}}{\left(1+\frac{1}{n}\right)^{n+1}} > \frac{1}{1+\frac{1}{n}} \Leftrightarrow \left(\frac{n^2+2n}{(n+1)^2}\right)^{n+1} > \frac{n}{n+1} \Leftrightarrow$$

$$\left(1 - \frac{1}{(n+1)^2}\right)^{n+1} > \frac{n}{n+1}. \tag{*}$$

However the last inequality in (*) is true, as it follows directly from (3-6), with $a = \frac{1}{n+1}$.

Finally $\frac{x_{n+1}}{x_n} > 1$, $\forall n \in \mathbb{N}$, or $x_{n+1} > x_n$, meaning that (x_n) is increasing.

Note: In a similar manner, one may show that the sequence $y_n = \left(1 + \frac{1}{n}\right)^{n+1}$ is decreasing, (see Problem (3-2)).

Example 3-3.

Show that the sequence $x_n = \left(1 + \frac{1}{n}\right)^n$ has an upper bound, while the sequence $y_n = \left(1 + \frac{1}{n}\right)^{n+1}$ has a lower bound.

Solution

In the previous example we proved that $(x_n) \nearrow$ while $(y_n) \searrow \cdot$
Also it is obvious that,

$$\left(1+\tfrac{1}{n}\right)^n < \left(1+\tfrac{1}{n}\right)^{n+1} \Leftrightarrow x_n < y_n, \ \forall n \in \mathbb{N}.$$

Therefore $x_n < y_n < y_{n-1} < y_{n-2} < \cdots < y_3 < y_2 < y_1 = 4$,
meaning that , $x_n < 4$, $\forall n \in \mathbb{N}$, i.e the number 4 is an upper bound for (x_n).

Similarly $2 = x_1 < x_2 < x_3 < \cdots < x_{n-2} < x_{n-1} < x_n < y_n$,, meaning that $2 < y_n$, $\forall n \in \mathbb{N}$ i.e. the number 2 is a lower bound for (y_n).

Example 3-4.
Investigate the monotonicity of the sequence x_n , defined by the recursive formula,

$$x_{n+1} = c + \frac{x_n}{x_n + c}, \quad x_1 = d > 0, \quad c > d > 0.$$

Solution
Let us consider the difference

$$x_{n+1} - x_n = c + \frac{x_n}{x_n+c} - c - \frac{x_{n-1}}{x_{n-1}+c} = \frac{c}{(x_n+c)\cdot(x_{n-1}+c)} \cdot (x_n - x_{n-1}). \quad (*)$$

Since $c > 0$ and $x_n > 0$, $\forall n \in \mathbb{N}$, the sign of the difference $(x_{n+1} - x_n)$ will be the same with the sign of the difference $(x_n - x_{n-1})$,for all values of n, i.e.

$$sign(x_{n+1} - x_n) = sign(x_n - x_{n-1}) = sign(x_{n-1} - x_{n-2}) = \cdots = sign(x_3 - x_2)$$
$$= sign(x_2 - x_1) = sign\left(c + \frac{x_1}{x_1 + c} - x_1\right) =$$
$$sign\left(c + \frac{d}{d+c} - d\right), \quad\quad\quad (**)$$

However, since $c > d > 0$, by assumption, the last term in (**) is a positive number, and following (**) backwards , $sign(x_{n+1} - x_n) = +$,meaning that
$$x_{n+1} - x_n > 0, \quad or \quad x_{n+1} > x_n, \quad \forall n \in \mathbb{N},$$
therefore (x_n) is an increasing sequence.

PROBLEMS

3-1) Show that the sequence $x_n = \dfrac{1}{n+1} + \dfrac{1}{n+2} + \dfrac{1}{n+3} + \cdots + \dfrac{1}{2n}$ is increasing.

Hint: Consider the difference $(x_{n+1} - x_n)$, and show that $x_{n+1} - x_n > 0$, $\forall n \in \mathbb{N}$.

3-2) Show that the sequence $y_n = \left(1 + \dfrac{1}{n}\right)^{n+1}$ is decreasing.

Hint: Consider the ratio $\dfrac{y_{n+1}}{y_n}$, and make use of the Bernoulli in equality (3-5).

3-3) Show that the sequence $y_n = \dfrac{2^n}{(n+1)!}$ is decreasing.

Hint: The reader is supposed to be familiar with **the symbol** $n!$ **(read n factorial)**, which is defined as $n! = 1 \cdot 2 \cdot 3 \cdots n$, $1! = 1, 0! = 1$.

Consider the ratio $\dfrac{y_{n+1}}{y_n}$, and simplify, taking into consideration that $(n+2)! = (n+1)! \cdot (n+2)$.

3-4) Show that the sequence $x_n = \dfrac{1}{1!} + \dfrac{1}{2!} + \dfrac{1}{3!} + \cdots + \dfrac{1}{n!}$ is increasing, while the sequence $y_n = \dfrac{1}{1!} + \dfrac{1}{2!} + \dfrac{1}{3!} + \cdots + \dfrac{1}{n!} + \dfrac{1}{n \cdot n!}$ is decreasing.

3-5) Determine if the given sequences are monotone, and classify them as increasing or decreasing.
a) $x_n = \dfrac{n^n}{n!}$, **b)** $y_n = \dfrac{n}{3^n}$, **c)** $w_n = \dfrac{2^n}{1+2^{2n}}$,

d) $a_n = \tan^{-1} n$, **e)** $b_n = n - \dfrac{1}{n}$, **f)** $c_n = \dfrac{3n-1}{4n+2}$.

(Answer: a) Increasing, **b)** Decreasing, **c)** Decreasing, **d)** Increasing, **e)** Increasing, **f)** Increasing).

Hint: For (d), $a_1 = \tan^{-1} 1 = \dfrac{\pi}{4}$, $a_2 = \tan^{-1} 2$, $\cdots$ As n increases, the corresponding terms a_n approach $\dfrac{\pi}{2}$. Remember that the function $\tan^{-1}(x)$ is increasing in the interval $0 < x < \dfrac{\pi}{2}$.

3-6) Show that the sequence $x_{n+1} = \sqrt{x_n + c}$, $x_1 = \sqrt{c}$, $c > 0$, is increasing. (In Example 2-5, we proved that the number $(c + 1)$, is an upper bound for (x_n) .

Hint: Consider the difference $(x_{n+1} - x_n)$ and show that
$sign(x_{n+1} - x_n) = sign(x_n - x_{n-1}) =\cdots= sign(x_2 - x_1) = +$

3-7) If (x_n) is increasing (decreasing), show that the sequence
$$y_n = \frac{x_1 + x_2 + x_3 + \cdots x_n}{n},$$
will also be increasing (decreasing).

3-8) Let the sequence (x_n) be defined by the recursive formula,
$$x_{n+1} = 1 + \frac{x_n}{1 + x_n} , \qquad x_1 = c > 0.$$

a) Show that ,

If $c > \frac{1+\sqrt5}{2}$, (x_n) is decreasing.

If $0 < c < \frac{1+\sqrt5}{2}$, (x_n) is increasing.

If $c = \frac{1+\sqrt5}{2}$, (x_n) is constant, $(x_1 = x_2 = x_3 =\cdots= c)$.

b) Show that the sequence (x_n) is bounded.

Hint: a) Consider the difference $x_{n+1} - x_n$, and show that
$$sign(x_{n+1} - x_n) = sign(x_n - x_{n-1}) =\cdots= sign(x_2 - x_1)$$
$$= sign\left(1 + \frac{c}{1 + c} - c\right).$$
b) Since all the terms of (x_n) are positive, it is easy to show that $1 < x_n < 2$, for all values of n, (Prove it).

3-9) Consider the sequence $x_{n+1} = \frac{1}{2}\left(x_n + \frac{1}{x_n}\right)$, $x_1 = c > 1.$
Show that the sequence (x_n) is decreasing and bounded.
Hint: Since $x_n > 0$, $\forall n \in \mathbb{N}$, $x_n + \frac{1}{x_n} \geq 2$ (why?).

3-10) Show that the sequence $y_n = (n + 1)^{\frac{1}{n}}$, is decreasing and bounded.

Hint: You may use equation (3-6).

3-11) Show that the sequence $x_{n+1} = 3 + \frac{2x_n}{3x_n+2}$, $x_1 = 3$, is increasing and bounded.

3-12) For what values of c is $x_n = \frac{2n+c}{2n+1}$ increasing?

3-13) If $x > 1$, show that the sequence $y_n = \sqrt[n]{x}$ is decreasing and bounded below.

3-14) Prove that the sequence $x_n = \frac{8n}{n+3}$ is bounded and increasing.

3-15) Prove that the sequence $y_n = \frac{5^n+3}{5^n}$ is bounded and decreasing.

3-16) Prove the two Bernoulli inequalities, (3-5) and (3-6), using Mathematical induction.

4. Subsequences.

Let $(x_n) = \{x_1, x_2, x_3, \cdots, x_n, x_{n+1}, \cdots\}$ be a sequence of real numbers.

Let also $\{r_1, r_2, r_3, \cdots, r_n, r_{n+1}, \cdots\}$ **be a strictly increasing sequence of positive integers**.

We now define a new sequence (y_n), having general term,

$$y_n = x_{r_n}, \text{ i.e. } \{y_1 = x_{r_1}, \ y_2 = x_{r_2}, \ y_3 = x_{r_3}, \dots, y_n = x_{r_n}, \dots \}.$$

$$(4\text{-}1)$$

The sequence (y_n) is called a subsequence of the sequence (x_n).

For example, if we take

$r_1 = 2, \ r_2 = 4, \ r_3 = 6, \ r_4 = 8, \cdots$, then the sequence (y_n) defined as,

$\{y_1 = x_2, \ y_2 = x_4, \ y_3 = x_6, \ y_4 = x_8, \cdots\}$ will be a subsequence of (x_n).

Another subsequence of (x_n) will be the sequence (w_n) defined as,

$\{w_1 = x_1, \ w_2 = x_3, \ w_3 = x_5, \ w_4 = x_7, \cdots\}$.

Theorem 4-1.

If a sequence (x_n) is increasing (decreasing) then any subsequence of (x_n) will be increasing (decreasing).

Theorem 4-2.

If a sequence (x_n) is bounded, then any subsequence of (x_n) will also be bounded, by the same lower and upper bound.

The proof is easy, (see Problems 4-1 and 4-2).

Example 4-1.

If $x_n = \left(1 + \frac{1}{n}\right)^n$ and $r_n = n^2$ $(n \in \mathbb{N})$, list the first five terms of the subsequence $y_n = x_{r_n}$.

Solution

The sequence $r_n = n^2$ is obviously an increasing sequence of positive integers, with $r_1 = 1, r_2 = 4, r_3 = 9, r_4 = 16, r_5 = 25, \cdots$

The first five terms of the subsequence $y_n = x_{r_n}$, will be,

$$y_1 = x_{r_1} = x_1 = \left(1 + \frac{1}{1}\right)^1 = 2,$$

$$y_2 = x_{r_2} = x_4 = \left(1 + \frac{1}{4}\right)^4 = \left(\frac{5}{4}\right)^4,$$

$$y_3 = x_{r_3} = x_9 = \left(1 + \frac{1}{9}\right)^9 = \left(\frac{10}{9}\right)^9,$$

$$y_4 = x_{r_4} = x_{16} = \left(1 + \frac{1}{16}\right)^{16} = \left(\frac{17}{16}\right)^{16},$$

$$y_5 = x_{r_5} = x_{25} = \left(1 + \frac{1}{25}\right)^{25} = \left(\frac{26}{25}\right)^{25}.$$

Example 4-2.

If $x_n = \frac{1}{n+5}$ and $r_n = 2n + 1$, $(n \in \mathbb{N})$, list the first five terms of the subsequence $y_n = x_{r_n}$.

Solution

The sequence $r_n = 2n + 1$, is an increasing sequence of positive integers with $r_1 = 3, r_2 = 5, r_3 = 7, r_4 = 9, r_5 = 11, \cdots$.

The first five terms of the subsequence $y_n = x_{r_n}$, will be,

$$y_1 = x_{r_1} = x_3 = \frac{1}{3+5} = \frac{1}{8},$$

$$y_2 = x_{r_2} = x_5 = \frac{1}{5+5} = \frac{1}{10},$$

$$y_3 = x_{r_3} = x_7 = \frac{1}{7+5} = \frac{1}{12},$$

$$y_4 = x_{r_4} = x_9 = \frac{1}{9+5} = \frac{1}{14},$$

$$y_5 = x_{r_5} = x_{11} = \frac{1}{11+5} = \frac{1}{16}.$$

Example 4-3.

If $x_n = \dfrac{\cos(n\pi)}{n^2}$, find the subsequences, $y_n = x_{2n}$ and $w_n = x_{2n+1}$.

Solution

The subsequence $y_n = x_{2n} = \dfrac{\cos(2n\pi)}{(2n)^2} = \dfrac{1}{(2n)^2}$, while $w_n = x_{2n+1} =$

$\dfrac{\cos((2n+1)\pi)}{(2n+1)^2} = \dfrac{(-1)}{(2n+1)^2}$.

PROBLEMS

4-1) Prove Theorem 4-1.

4-2) Prove Theorem 4-2.

4-3) If $x_n = \dfrac{n}{n^2+2}$ and $r_n = n^2 + 1$, list the first five terms of the subsequence $y_n = x_{r_n}$.

(**Answer:** $y_1 = \dfrac{2}{6}$, $y_2 = \dfrac{5}{27}$, $y_3 = \dfrac{10}{102}$, $y_4 = \dfrac{17}{291}$, $y_5 = \dfrac{26}{678}$).

4-4) If $x_n = \dfrac{n^2}{n^3+1}$ and $r_n = (n+1)^2$ list the first five terms of the subsequence $y_n = x_{r_n}$.

4-5) Show that the subsequence (y_n) in Problem 4-4 is decreasing and bounded.

4-6) Show that the two subsequences (y_n) and (w_n), in Example 4-3, are absolutely bounded.

4-7) If $x_n = \tan^{-1} n$, show that the sequence (x_n) is increasing and bounded, and then show that the subsequence $y_n = x_{2n+1}$ has the same properties.

4-8) If $x_n = \cos(n\pi)$, show that $y_n = x_{2n} = 1$ and $w_n = x_{2n+1} = -1$.

5. Null Sequences.

Let us consider the sequence $a_n = \frac{1}{n}$, $n \in \mathbb{N}$.

The first few terms of the sequence (a_n), are,

$$\{a_1 = \frac{1}{1} = 1, \ a_2 = \frac{1}{2}, \ a_3 = \frac{1}{3}, \cdots, a_{100} = \frac{1}{100}, \cdots, \ a_{1000} = \frac{1}{1000}, \cdots\}.$$

We note that **no term of (a_n) is equal to zero**. However, as the index n increases, the corresponding terms of the sequence decrease and approach zero, or stated differently, as the index n increases, the corresponding terms of the sequence become **arbitrarily small**.

What does the term **"arbitrarily small"** mean? Let us take a very small **positive** number, for example the number 10^{-20}. Then form a certain value of the index n on, all the corresponding terms of the sequence (a_n), will be smaller than 10^{-20}. Indeed, it is easily verified that $a_n = \frac{1}{n} < 10^{-20}$ for **all** values of $n > 10^{20}$.

Had we chosen an even smaller **positive** number, for example the number $10^{-50} < 10^{-20}$, then again, from a certain value of the index n on, all the corresponding terms of the sequence (a_n) would be smaller than 10^{-50}. Indeed, $a_n = \frac{1}{n} < 10^{-50}$ for **all** values of $n > 10^{50}$.

Continuing this mental process we may say that if ε is any small positive number, however small this number might be, (**arbitrarily small**), then from a certain value of the index n on, all the corresponding terms of the sequence $a_n = \frac{1}{n}$, will be smaller than ε. Indeed, **for any arbitrarily small positive number ε, $a_n = \frac{1}{n} <$** ε, for all values of $n > \frac{1}{\varepsilon}$.

In this case we say that **the sequence (a_n) tends to zero**, or that **the limit of (a_n) is zero**, as n approaches infinity, and we write,

$$\lim_{n\to\infty} a_n = 0 \quad or \quad a_n \xrightarrow[n\to\infty]{} 0. \tag{5-1}$$

Equivalent expressions are, **the sequence** (a_n) **converges to zero** or that (a_n) **is a null sequence**.

Following this brief introduction, we are now in a position to give a more rigorous definition for null sequences.

Definition 5-1: A sequence (a_n) is called **a null sequence**, if given **any arbitrarily small positive number** ε, the terms of the sequence, from **a certain value of the index** n **on**, become in **absolute value** smaller than ε.

In symbols,

$$\lim_{n\to\infty} a_n = 0 \iff \forall \varepsilon > 0, \ \exists N = N(\varepsilon): \ \forall n > N \implies |a_n| < \varepsilon. \qquad (5\text{-}2)$$

The key word in (5-2) is $\forall \, \varepsilon > 0$, **(for every** $\varepsilon > 0$).

The sequence (a_n) will be a null sequence, if (5-2) is satisfied **for every** $\varepsilon > 0$.

The notation $\exists \, N = N(\varepsilon)$ means that there exists an index N, which **in general depends on the arbitrarily chosen** $\varepsilon > 0$, such that all the terms of the sequence with index greater than $N = N(\varepsilon)$ will satisfy $|a_n| < \varepsilon$, i.e.

$$|a_{N+1}| < \varepsilon, \ |a_{N+2}| < \varepsilon, \ |a_{N+3}| < \varepsilon, \ |a_{N+4}| < \varepsilon, \ |a_{N+5}| < \varepsilon, \cdots \qquad (5\text{-}3)$$

Summarizing, if one wants to show that a given sequence (a_n) is a null sequence, then

a) An arbitrarily small, positive number ε, must be chosen,

b) An index N, (depending on ε, $N = N(\varepsilon)$) must be found, and finally

c) Show that for all indices $n > N$, the corresponding terms of the sequence satisfy $|a_n| < \varepsilon$.

In general **the smaller the** ε **is, the greater** $N = N(\varepsilon)$ **will be**.

Definition 5-2: Let ℓ be a given real number and ε be any positive number. The totality of numbers x, for which $|x - \ell| < \varepsilon$, is called an ε neighborhood of ℓ.

We note that

$$|x - \ell| < \varepsilon \Leftrightarrow -\varepsilon < x - \ell < \varepsilon \Leftrightarrow \ell - \varepsilon < x < \ell + \varepsilon. \qquad (5\text{-}4)$$

In this case we say that x **lies within an** ε **neighborhood of the number** ℓ.

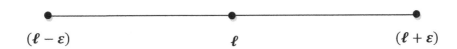

Fig. 5-1: An ε neighborhood of ℓ.

In particular, if $\ell = 0$, then the inequality $|x| < \varepsilon \Leftrightarrow -\varepsilon < x < \varepsilon$, **defines an** ε **neighborhood of zero**.

In the light of (5-2), (5-4) and the definition 5-2, we may say, in loose terms that, **if a given sequence (a_n) is a null sequence, then every ε neighborhood of zero, contains an infinite number of terms of (a_n).**

(**All terms** $a_{N+1}, a_{N+2}, a_{N+3}, a_{N+4}, a_{N+5}, \cdots$, lie in an ε neighborhood of zero).

Example 5-1.

Show that the sequence $x_n = \dfrac{(-1)^n}{n}$, $n \in \mathbb{N}$ is a null sequence.

Solution

Let $\varepsilon > 0$ be any arbitrarily small positive number. Then $|x_n| = \dfrac{1}{n} < \varepsilon$, provided that $n > \dfrac{1}{\varepsilon}$.

So if we choose $N = N(\varepsilon) = \left[\dfrac{1}{\varepsilon}\right] + 1$, (**the integer part of** $\dfrac{1}{\varepsilon}$ **plus one**), then, the infinite number of terms $\{x_{N+1}, x_{N+2}, x_{N+3}, \cdots\}$ will, in absolute value, be less than ε, meaning that $\lim_{n \to \infty} x_n = 0$.

Note: If x is any real number, then **the integer part of x, (symbol $[x]$), is the greatest integer not exceeding x**. For example $[3.7] = 3$, $[\sqrt{2}] = 1$, $[-6.5] = -7$, etc.

In general $[x] \leq x < [x] + 1$. $\hspace{4cm}$ (5-5)

Example 5-2.

Show that the sequence $y_n = \frac{1}{n^2}$ $n \in \mathbb{N}$ is a null sequence.

Solution

Let $\varepsilon > 0$ be any arbitrarily small positive number. Then $|y_n| = \frac{1}{n^2} < \varepsilon$, provided that $n^2 > \frac{1}{\varepsilon}$,or equivalently, $n > \frac{1}{\sqrt{\varepsilon}}$.

So if we choose $N = N(\varepsilon) = \left[\frac{1}{\sqrt{\varepsilon}}\right] + 1$, all the terms $y_{N+1}, y_{N+2}, y_{N+3}, \cdots$, will, in absolute value be less than ε, meaning that $\lim_{n \to \infty} y_n = 0$.

Note: In any completely similar manner, one may show that the sequences $a_n = \frac{1}{n^3}$, $b_n = \frac{1}{n^4}$, $c_n = \frac{1}{n^5}$, $d_n = \frac{1}{n^6}$, $\cdots$, are all null sequences. Also the sequences $x_n = \frac{1}{\sqrt{n}}$, $y_n = \frac{1}{\sqrt[3]{n}}$, $z_n = \frac{1}{\sqrt[4]{n}}$, etc. are all null sequences.

A very general remark:

In practice one rarely uses the Definition 5-1, in order to show that a given sequence is a null sequence. Instead, the usual procedure is to prove some general Theorems first, by means of which, one may, relatively easily, investigate if a given sequence is a null sequence i.e. if it converges to zero, (or to any other number).

PROBLEMS

5-1) Show that the following are null sequences,

a) $x_n = \frac{1}{n^5}$, $\hspace{0.5cm}$ **b)** $y_n = \frac{1}{\sqrt{n}}$, $\hspace{0.5cm}$ **c)** $w_n = \frac{1}{n+1}$.

5-2) Show that $a_n = \frac{\cos(n\pi)}{n^2}$, is a null sequence.

5-3) Show that $c_n = \frac{n}{n^2+1}$, is a null sequence.

5-4) If (x_n) is bounded and (y_n) is null, show that the sequence $(x_n \cdot y_n)$ is null.

5-5) Show that the following are null sequences,

    a) $x_n = \dfrac{1}{n^{\frac{3}{2}}}$,   **b)** $y_n = \dfrac{1}{7n^5}$.

5-6) Making use of Problem 5-4 show that the sequence $x_n = \dfrac{\sin(3\sqrt{n}+2)}{n}$ is null.

5-7) Show that the subsequence of any null sequence is a null sequence, as well.

5-8) If x_n is a null sequence and c is any fixed real number $\neq 0$, show that the sequence (cx_n) is null.

5-9) Find the integer part of the numbers $(\sqrt[3]{-12})$ and $\left(\dfrac{3}{5} - \dfrac{16}{11}\right)$.

(Answer: $-3,-1$).

5-10) If x is any real number, show that $0 \le x - [x] < 1$.

6. Convergent Sequences.

In this chapter we shall give the definition of a sequence, converging not to zero, but to another **finite, real number ℓ**.

This definition, however, relies heavily on the definition of null sequences.

Definition 6-1: We say that a sequence (x_n) converges to a finite number ℓ, if the sequence $(x_n - \ell)$ is a null sequence. **The number ℓ is called the limit of** (x_n). In symbols,

$$\lim_{n\to\infty} x_n = \ell \Leftrightarrow \forall\, \varepsilon > 0, \exists\, N = N(\varepsilon): \forall\, n > N \implies |(x_n - \ell)| < \varepsilon. (6\text{-}1)$$

Making use of (5-4) and the definition (5-2), we may say, that if a sequence (x_n) converges to a finite limit ℓ, then **every** ε neighborhood of ℓ, contains an infinite number of terms of (x_n).

(All the terms $x_{N+1}, x_{N+2}, x_{N+3}, x_{N+4}, \ldots$ lie in an ε neighborhood of ℓ, i.e. in the interval $(\ell - \varepsilon, \ell - \varepsilon)$).

A sequence need not have a finite limit, (not necessarily).

For example the sequences $x_n = n^2$, $(x_1 = 1, x_2 = 4, x_3 = 9, \cdots)$ and $y_n = (-1)^n n$, $(y_1 = -1, y_2 = 2, y_3 = -3, y_4 = 4, \cdots)$ have **no finite limits**.

However **if a sequence has a finite limit, this limit is unique.** (This important theorem will be proved shortly, in Chapter 10, Theorem 10-1).

If $\lim_{n\to\infty} x_n = \ell$, we say that **the sequence (x_n) is convergent to the limit ℓ**, or that **its terms tend to the limit ℓ, as $n \to \infty$**, and we may write $x_n \xrightarrow[n\to\infty]{} \ell$.

Example 6-1.

If $x_n = \dfrac{n+1}{n}$, show that $\lim_{n\to\infty} x_n = 1$.

Solution

It suffices to show that the sequence $y_n = x_n - 1$, is a null sequence (Definition 6-1).

Indeed, $y_n = x_n - 1 = \frac{n+1}{n} - 1 = 1 + \frac{1}{n} - 1 = \frac{1}{n}$, and since the sequence $\left(y_n = \frac{1}{n}\right)$ is a null sequence, the $\lim_{n\to\infty} x_n = 1$.

Example 6-2.

If $x_n = \frac{3n^2+5}{2n^2+1}$ show that $\lim_{n\to\infty} x_n = \frac{3}{2}$.

Solution

It suffices to show that the sequence $y_n = x_n - \frac{3}{2}$ is a null sequence.

Indeed, $y_n - \frac{3}{2} = \frac{3n^2+5}{2n^2+1} - \frac{3}{2} = \frac{7}{2(2n^2+1)}$, and since the $\lim_{n\to\infty} \frac{7}{2(2n^2+1)} = 0$ (why?), the $\lim_{n\to\infty} \frac{3n^2+5}{2n^2+1} = \frac{3}{2}$.

PROBLEMS

6-1) Show that $\lim_{n\to\infty} \frac{n^2+1}{n^2} = 1$.

6-2) Show that $\lim_{n\to\infty} \frac{n^k+1}{n^k} = 1$, where k is any fixed, positive integer.

6-3) Show that $\lim_{n\to\infty} \frac{5n^4+8}{7n^4} = \frac{5}{7}$.

6-4) Show that $\lim_{n\to\infty} \frac{2\sqrt{n}}{3+5\sqrt{n}} = \frac{2}{5}$.

6-5) Show that $\lim_{n\to\infty} \frac{e^n+2}{e^n-2} = 1$.

A very general remark: In the Examples and the Problems of this chapter we use the Definition 6-1, in order to show that the given sequence converges to a given finite limit. **In the Definition 6-1, the limit ℓ of the sequence(x_n), is assumed to be known in advance.** However, in most cases, the limit of a given sequence (x_n), assuming it exists, **is not known in advance.** As a matter of fact the main problem in sequences is, given a sequence (x_n) to find its limit (assuming it exists). In practice, the limit of a given sequence is determined with the aid of certain rules and Theorems, to be developed in the chapters to follow.

7. Divergent Sequences.

There are sequences, like $x_n = n$ or $y_n = n^2$ or $w_n = 7n^3 + 5$, etc, which, as n increases (and approaches infinity), the corresponding terms of the sequence become **arbitrarily large**, meaning that **from a given stage on, all the terms of the sequence will exceed any given positive number M, no matter how large M might be.**

Intuitively, we understand, that **if M is any given large positive number**, then the terms of the sequence $(y_n = n^2)$, i.e.
$$\{y_1 = 1^2,\ y_2 = 2^2,\ y_3 = 3^2,\cdots, y_{1000} = 1000^2,\cdots\}$$
from a certain point on, will exceed any arbitrarily chosen $M > 0$, **no matter how large M might be.**

In this case we say that **the sequence $(y_n = n^2)$ is divergent, or that it diverges to infinity**, and we write

$$\lim_{n\to\infty} y_n = +\infty \ \text{ or } \ y_n \xrightarrow[n\to\infty]{} +\infty. \tag{7-1}$$

More precisely, we have the following definitions:

Definition 7-1: We say that **a sequence (x_n) tends to $+\infty$ or diverges to $+\infty$** if, given an **arbitrary** positive number $M > 0$, it is possible to find a positive integer $N = N(M)$ such that $x_n > M$ for every $n > N = N(M)$.
In symbols,

$$\lim_{n\to\infty} x_n = +\infty \iff \forall\, M > 0,\ \exists\, N = N(M)\colon \forall\, n > N \implies x_n > M. \tag{7-2}$$

In a similar manner, we may easily give the definition of a sequence approaching or tending to $-\infty$, as $n \to \infty$.

Definition 7-2: We say that **a sequence (y_n) tends to $-\infty$ or diverges to $-\infty$** if, given an **arbitrary** positive number $M > 0$, it is possible to find a positive integer $N = N(M)$ such that $y_n < -M$ for every $n > N = N(M)$.
In symbols,

$\lim_{n\to\infty} y_n = -\infty \Leftrightarrow \forall M > 0, \exists N = N(M): \forall n > N \Rightarrow y_n < -M.$ (7-3)

For example the sequences $b_n = -n^2$, $c_n = -n^3$ and $d_n = -n^4$, $n\epsilon\mathbb{N}$, approach $-\infty$, as $n \to \infty$. There are sequences, which are neither convergent nor divergent, for example the sequence $x_n = (-1)^n n$.

Example 7-1.

If $x_n = n^2$, show that $\lim_{n\to\infty} x_n = +\infty$.

Solution

Let M be any given positive number,(**arbitrarily chosen**). Then, it suffices to find a positive integer $N = N(M)$ such that for all $n > N$, the corresponding terms of the sequence $x_n > M$.
Indeed, (since $x_n > 0$, $\forall n \in \mathbb{N}$), $x_n > M \Leftrightarrow n^2 > M \Leftrightarrow n > \sqrt{M}$. (*)

So if we choose $N = \left[\sqrt{M}\right] + 1$, then $\forall n > N \Rightarrow x_n > M$, (from (*)), meaning that $\lim_{n\to\infty} x_n = +\infty$.

Example 7-2.

If $\lim_{n\to\infty} x_n = +\infty$ and $y_n = -x_n$, show that $\lim_{n\to\infty} y_n = -\infty$.

Solution

Since $\lim_{n\to\infty} x_n = +\infty$,

$\forall M > 0, \exists N = N(M): \forall n > N \Rightarrow x_n > M,$ (*)

or, since $y_n = -x_n$,

$\forall M > 0, \exists N = N(M): \forall n > N \Rightarrow -y_n > M \Leftrightarrow$

$\forall M > 0, \exists N = N(M): \forall n > N \Rightarrow y_n < -M,$

meaning that $\lim_{n\to\infty} y_n = -\infty$, (see (7-3)).

PROBLEMS

7-1) If $x_n = n^3$, $n \in \mathbb{N}$, show that $\lim_{n\to\infty} x_n = +\infty$.

7-2) If $y_n = -(n+1)$, $n \in \mathbb{N}$, show that $\lim_{n\to\infty} y_n = -\infty$, while $\lim_{n\to\infty}(y_n^2) = +\infty$.

7-3) If $x_n = -n^3$ and $y_n = (n+1)^3$, show that $\lim_{n\to\infty}(x_n + y_n) = +\infty$.

7-4) If $\lim x_n = +\infty$ and $\lim y_n = -\infty$, does the sequence $(x_n + y_n)$ always diverge? Justify your answer, by considering various Examples.

7-5) If $\lim_{n\to\infty} x_n = +\infty$ and c is any constant, show that $\lim_{n\to\infty}(c + x_n) = +\infty$.

7-6) If $\lim_{n\to\infty} x_n = +\infty$ and c is any constant $(c \neq 0)$, show that

$$\lim_{n\to\infty}(c\,x_n) = \begin{cases} +\infty & if \ \ c > 0 \\ -\infty & if \ \ c < 0 \end{cases}.$$

7-7) If $x_n > y_n$, $\forall n \in \mathbb{N}$ and $\lim_{n\to\infty} y_n = +\infty$ then $\lim_{n\to\infty} x_n = +\infty$.

7-8) If the terms of a sequence (x_n) are positive numbers, and $\lim_{n\to\infty} x_n = +\infty$, show that $\lim_{n\to\infty} \sqrt{x_n} = +\infty$, and $\lim_{n\to\infty} \sqrt[3]{x_n} = +\infty$.

7-9) Show that if $\lim_{n\to\infty} x_n = +\infty$, then the $\lim_{n\to\infty}\left(\frac{1}{x_n}\right) = 0$.
(Assume that $x_n \neq 0$, $\forall n \in \mathbb{N}$).

7-10) Show that $\lim_{n\to\infty}(2n^2 + 5n + 3) = +\infty$ and that $\lim_{n\to\infty}(n^3 - n^2) = +\infty$.

7-11) Show that $\lim_{n\to\infty}(-2n^2 + 5n + 3) = -\infty$.

8. Basic Properties for Null and Divergent Sequences.

Given two sequences (x_n) and (y_n) we may form some new sequences, for example, $(x_n + y_n)$, $(x_n - y_n)$, $(x_n \cdot y_n)$, $\left(\frac{x_n}{y_n}\right)$ by adding, subtracting, multiplying or dividing term wise the given sequences. We may also form the sequence $(|x_n|)$ by taking the absolute values of the terms of (x_n), the sequence $(c \cdot x_n)$, where c is a given constant number, the sequence $(c \cdot x_n + d \cdot y_n)$, where c and d are given constants, the sequences $(x_n{}^k)$ or $\left(\sqrt[k]{y_n}\right)$ where k is any fixed positive integer etc.

We can now state some simple, fundamental rules and properties applying to null and divergent sequences.

For brevity and notation economy, we may sometimes write, $\lim x_n$ or $\lim y_n$ instead of $\lim_{n\to\infty} x_n$ or $\lim_{n\to\infty} y_n$, respectively. **The limit of any sequence is understood, in the sense that $n \to +\infty$.**

a) Assuming that (x_n) is a null sequence **($\lim x_n$=0)**, and that **k is any fixed positive integer**, then,

1) $\lim x_n{}^k = 0,$ (8-1)

2) $\lim \sqrt[k]{x_n} = 0,$ (assuming $x_n \geq 0$), (8-2)

3) $\lim(c \cdot x_n) = 0,$ (c is arbitrary constant $\neq 0$), (8-3)

4) $\lim|x_n| = 0.$ (8-4)

b) Assuming that (x_n) is a null sequence, **($\lim x_n = 0$)**, (y_n) and (w_n) are divergent to $+\infty$ sequences, **($\lim y_n = +\infty$)**, **($\lim w_n = +\infty$)**, and that **k is any fixed positive integer**, then ,

5) $\lim\left(\frac{1}{x_n}\right) = +\infty,$ (assuming $x_n > 0$),   (8-5)

6) $\lim\left(\frac{1}{y_n}\right) = 0,$ (assuming $y_n \neq 0$),   (8-6)

7) $\lim(y_n + w_n) = +\infty,$ (8-7)

8) $\lim(y_n \cdot w_n) = +\infty,$ (8-8)

9) $\lim y_n{}^k = +\infty,$ (8-9)

10) $\lim \sqrt[k]{y_n} = +\infty,$ (assuming $y_n > 0$),   (8-10)

11) $\lim\left(\frac{x_n}{y_n}\right) = 0,$ (assuming $y_n \neq 0$),   (8-11)

12) $\lim(c \cdot y_n) = \begin{cases} +\infty & \text{if } c \text{ is a positive constant} \\ -\infty & \text{if } c \text{ is a negative constant} \end{cases}.$ (8-12)

Further to these simple properties, we may state the following important Theorems, used quite often in practice, in order to evaluate the limit of a given sequence.

Theorem 8-1.

Let x be a fixed real number.

Then the $\lim x^n = \begin{cases} +\infty & \text{if } x > 1 \\ 0 & \text{if } |x| < 1 \end{cases}.$ (8-13)

(For a proof, see Example 8-4).

Theorem 8-2.

If x is a fixed real number > 1, then $\lim\left(\frac{x^n}{n}\right) = +\infty$ (8-14)

(For a proof, see Problem 8-4).

Corollary 1: If $x > 1$, then $\lim\left(\frac{n}{x^n}\right) = 0.$ (8-15)

Corollary 2: If $0 < |x| < 1$, then $\lim(nx^n) = 0.$ (8-16)

Example 8-1.
Show property (8-1).
Solution

Since (x_n) is a null sequence, given any $\varepsilon > 0$, there exists a positive number $N = N(\varepsilon)$, such that $\forall\, n > N \Rightarrow |x_n| < \sqrt[k]{\varepsilon}$, meaning that $\forall\, n > N \Rightarrow |x_n|^k = |x_n{}^k| < \varepsilon$, which in turns means that (x_n) is a null sequence.

Example 8-2.
Show property (8-5).
Solution

Let $M > 0$ be any fixed, arbitrarily large positive number. Since (x_n) is a null sequence, given any $\varepsilon > 0$, there exists a positive number $N = N(\varepsilon)$, such that $\forall\, n > N \Rightarrow |x_n| < \varepsilon$, or $\quad 0 < x_n < \varepsilon$, since $x_n > 0$. $\qquad$ (*)

If we choose $\varepsilon = \dfrac{1}{M}$, (the larger the M, the smaller the ε will be), from (*) one obtains, $\forall\, n > N \Rightarrow 0 < x_n < \dfrac{1}{M}$ or, $\forall\, n > N \Rightarrow \dfrac{1}{x_n} > M$, which according to the (7-2), implies that $\lim \dfrac{1}{x_n} = +\infty$, and the proof is completed.

Example 8-3.
Show property (8-7).
Solution

Since $\lim y_n = +\infty$, given any arbitrary positive number $\dfrac{M}{2}$, there exists a positive integer $N_1 = N_1(\frac{M}{2})$, such that $\forall\, n > N_1 \Rightarrow y_n > \dfrac{M}{2}$. $\quad$ (*)

Similarly, since $\lim w_n = +\infty$, there exists a positive integer $N_2 = N_2(\frac{M}{2})$, such that $\forall\, n > N_2 \Rightarrow w_n > \dfrac{M}{2}$. $\qquad$ (**)

Let $N = \max(N_1, N_2)$. **For $n > N$, (*) and (**) are satisfied simultaneously**, and adding together yields,

$$\forall\, n > N \Rightarrow y_n + w_n > M,$$

meaning that $\lim(y_n + w_n) = +\infty$, (since M is an arbitrary positive number), and the proof is therefore completed.

Example 8-4.

Prove Theorem (8-1).

Solution

a) Let $x > 1$. Then $x - 1 > 0$, and $x = 1 + (x - 1)$, from which applying **Bernoulli's inequality** (3-5), one obtains,
$$x^n = \{1 + (x - 1)\}^n > 1 + n(x - 1), \quad n = 2,3,4,\cdots \qquad (*)$$
Since $1 + n(x - 1) \to +\infty$ as $n \to +\infty$ (why?), and $\{1 + (x - 1)\}^n > 1 + n(x - 1)$, the sequence $x^n = \{1 + (x - 1)\}^n \to +\infty$, as well, and this completes the proof. (See also Problem 7-7).

b) If $x = 0$, then obviously $\lim x^n = 0$. Let's now assume that $|x| < 1$, $(x \neq 0)$, or $\frac{1}{|x|} > 1$. Then according to part (a),
$$\lim\left(\frac{1}{|x|}\right)^n = \lim\left(\frac{1}{|x|^n}\right) = +\infty \text{ or } \lim|x|^n = 0, \text{ by virtue of (8-6), and}$$
finally $\lim x^n = 0$. (It is very easy to show that **if $\lim|x_n| = 0$, then $\lim x_n = 0$**, try it!).

Example 8-5.

Show that the sequence $x_n = \sqrt{3n + 2} - \sqrt{3n + 1}$ is null.

Solution

As $n \to \infty$, $(3n + 2) \to +\infty$ and $(3n + 1) \to +\infty$, and by property (8-10), $\sqrt{3n + 2} \to +\infty$ and $\sqrt{3n + 1} \to +\infty$. Formally we may say that, as $n \to \infty$ $x_n = \infty - \infty$, which is **an indeterminate form**, i.e. it could be anything. For the **proper evaluation of the $\lim x_n$,** we proceed as follows:

$$x_n = \sqrt{3n + 2} - \sqrt{3n + 1} = \frac{(\sqrt{3n+2}-\sqrt{3n+1})(\sqrt{3n+2}+\sqrt{3n+1})}{\sqrt{3n+2}+\sqrt{3n+1}} = \frac{(\sqrt{3n+2})^2-(\sqrt{3n+1})^2}{\sqrt{3n+2}+\sqrt{3n+1}} =$$

$$\frac{(3n+2)-(3n+1)}{\sqrt{3n+2}+\sqrt{3n+1}} = \frac{1}{\sqrt{3n+2}+\sqrt{3n+1}}.$$

As $n \to \infty, (\sqrt{3n+2} + \sqrt{3n+1}) \to +\infty$ (from (8-7)), therefore $\lim x_n = 0$, (from (8-6)), and the proof is completed.

PROBLEMS

8-1) Show properties (8-2) and (8-4).

8-2) Show properties (8-8) and (8-10).

8-3) Show properties (8-11) and (8-12).

8-4) Prove Theorem 8-2.

Hint: Since $x > 1$, we may set, $x = 1 + p$, where $p > 0$.

Making use of the Binomial Theorem,

$$x^n = (1+p)^n = 1 + n \cdot p + \frac{n(n-1)}{1 \cdot 2}p^2 + \frac{n(n-1)(n-2)}{1 \cdot 2 \cdot 3}p^3 + \cdots + p^n \Rightarrow$$

$$x^n = (1+p)^n > \frac{n(n-1)}{1 \cdot 2}p^2 \Rightarrow \frac{x^n}{n} > \frac{n-1}{2}p^2.$$

Notice that as $n \to \infty, \frac{n-1}{2}p^2 \to +\infty$, etc.

8-5) Prove Corollary 1 and Corollary 2.

8-6) If $x_n = (\sqrt{5n+7} - \sqrt{5n+1})$, show that $\lim x_n = 0$.

8-7) If (x_n) is bounded and $\lim y_n = +\infty$, show that $\lim \frac{x_n}{y_n} = 0$.

8-8) If $b_n = \frac{\sin(3n^2+5)}{n^4+10}$, show that $\lim b_n = 0$.

Hint: Make use of Problem 8-7. Note that the sequence $x_n = \sin(3n^2 + 5)$ is bounded, while the sequence $y_n = n^4 + 10$, diverges to $+\infty$.

8-9) If $x_n = \dfrac{n^2}{n^3+\sqrt{n^6+1}}$, show that $\lim x_n = 0$.

Hint: $x_n = \dfrac{1}{n+\sqrt{n^2+\frac{1}{n^4}}}$, and as $n \to \infty$, the term $\sqrt{n^2+\dfrac{1}{n^4}} \to +\infty$, then apply (8-7)

and (8-6).

8-10) Show that the sequences $x_n = \dfrac{n}{2^n}$ and $y_n = n\left(\sin\left(\frac{1}{2}\right)\right)^n$, are null

sequences.

Hint: Apply (8-15) and (8-16).

8-11) If $\lim x_n = 0$ and $\lim y_n = +\infty$, show that $\lim(x_n + y_n) = +\infty$. As an

application, show that $\lim \dfrac{2n}{\sqrt{16n^4+3}+4n^2} = 0$.

Hint: $\dfrac{2n}{\sqrt{16n^4+3}+4n^2} = \dfrac{1}{\sqrt{4n^2+\frac{3}{4n^2}}+2n}$, $\lim(4n^2) = +\infty$, $\lim\left(\dfrac{3}{4n^2}\right) = 0$, etc.

9. Basic Properties of Convergent Sequences.

Let (x_n) and (y_n) be two convergent sequences, having limits the finite numbers ℓ and m, respectively, i.e.

$$\lim x_n = \ell, \quad \text{and} \quad \lim y_n = m. \tag{9-1}$$

Then,

1) $\lim(x_n + y_n) = \lim x_n + \lim y_n = \ell + m,$ $\qquad$ (9-2)

2) $\lim(x_n - y_n) = \lim x_n - \lim y_n = \ell - m,$ $\qquad$ (9-3)

3) $\lim(c \cdot x_n + d \cdot y_n) = c \cdot \lim x_n + d \cdot \lim y_n = c \cdot \ell + d \cdot m,$ $\quad$ (9-4)
   where **c and d are constants, i.e. independent of n quantities.**

4) $\lim(x_n \cdot y_n) = \lim x_n \cdot \lim y_n = \ell \cdot m,$ $\qquad$ (9-5)

5) $\lim(x_n{}^k) = (\lim x_n)^k = \ell^k, \quad k = 1,2,3,\cdots$ $\qquad$ (9-6)

6) $\lim\left(\dfrac{x_n}{y_n}\right) = \dfrac{\lim x_n}{\lim y_n} = \dfrac{\ell}{m},$ $\quad$ (provided $m \neq 0$), $\qquad$ (9-7)

7) If all the terms of (x_n) are positive and the $\lim x_n = \ell \geq 0$, then
   $$\lim\left(\sqrt[k]{x_n}\right) = \sqrt[k]{\lim x_n} = \sqrt[k]{\ell}, \quad k = 1,2,3,\cdots \tag{9-8}$$

8) For the sequence (x_n) in **(7)**,
   $$\lim\left(x_n{}^{\frac{k}{p}}\right) = (\lim x_n)^{\frac{k}{p}} = (\ell)^{\frac{k}{p}}, \tag{9-9}$$
   where k and p are fixed positive integers.

9) $\lim|x_n| = |\lim x_n| = |\ell|,$ $\qquad$ (9-10)

10) If $\lim w_n = \pm\infty$, then $\lim\left(\dfrac{x_n}{w_n}\right) = 0.$ $\qquad$ (9-11)

These are the basic, fundamental properties used in practice, in order to evaluate the limits of various sequences. The proof of all the properties relies heavily on the definition of the limit, (see (6-1)). Some of these properties are proved in the following Examples, while some others are left, with appropriate hints, to be proved by the reader, in the Problems, at the end of the chapter.

Important Note: Properties (1), (2), (3) and (4) are valid for any **finite** number of sequences. For example,

$$\lim(2a_n - 3b_n + 7c_n - 5d_n) = 2\lim a_n - 3\lim b_n + 7\lim c_n - 5\lim d_n, \text{ and}$$
$$\lim(a_n \cdot b_n \cdot c_n \cdot d_n) = \lim a_n \cdot \lim b_n \cdot \lim c_n \cdot \lim d_n, \text{ etc.}$$

If the number of terms is not fixed, i.e. independent of n, but on the contrary depends on n, then the aforementioned properties cannot be applied. As an example, let us consider the sequence $x_n = \dfrac{1}{n+1} + \dfrac{1}{n+2} + \dfrac{1}{n+3} + \cdots + \dfrac{1}{2n}$.

The number of summands is n. As $n \to \infty$, the number of summands $\to \infty$, while each term $\to 0$. This is a classic example of a "**0 · ∞**" situation, which could be anything. For the **proper evaluation** of such limits special techniques are required. **Some of these techniques are developed in Chapter 16.**

The following Examples, will clarify how the aforementioned properties and rules, are applied in practice, in order to find the limit of a given sequence.

Example 9-1.

Show equation (9-2).

Solution

It suffices to show that

$$\forall\, \varepsilon > 0 \quad \exists\, N = N(\varepsilon): \;\; \forall n > N \Longrightarrow |x_n + y_n - (\ell + m)| < \varepsilon \quad \text{(see (6-1))}.$$

$$|x_n + y_n - (\ell + m)| = |(x_n - \ell) + (y_n - m)| \le |x_n - \ell| + |y_n - m|. \qquad (*)$$

Since, by assumption $\lim x_n = \ell$ and $\lim y_n = m$, given any arbitrary $\varepsilon > 0$, we can find two integers N_1 and N_2, such that

$$|x_n - \ell| < \frac{\varepsilon}{2}, \quad \forall\, n > N_1 \text{ and} \qquad\qquad (**)$$

$$|y_n - m| < \frac{\varepsilon}{2}, \quad \forall\, n > N_2. \qquad\qquad (***)$$

If we set $N = \max(N_1, N_2)$, then for any $n > N$, inequalities $(**)$ and $(***)$ are **satisfied simultaneously**, therefore adding them together, yields,

$$|x_n - \ell| + |y_n - m| < \varepsilon, \quad \forall\, n > N,$$

and taking $(*)$ into account, we have,

$$|(x_n + y_n) - (\ell + m)| < \varepsilon, \quad \forall n > N,$$

Showing thus that $\lim(x_n + y_n) = \ell + m = \lim x_n + \lim y_n$,
and this completes the proof.

Example 9-2.

Show that every convergent sequence (x_n) is bounded. (This is actually an **important Theorem, about convergent sequences,** see Theorem 10-2).

Solution

Let us assume that (x_n) converges to a finite limit ℓ, i.e. $\lim x_n = \ell$. Then,

$$|x_n| = |(x_n - \ell) + \ell| \le |x_n - \ell| + |\ell|. \tag{*}$$

Given an arbitrary $\varepsilon > 0$, there exists a positive integer $N = N(\varepsilon)$, such that

$$\forall n > N \implies |x_n - \ell| < \varepsilon, \text{ (see (6-1)), which when combined with } (*), \text{ yields,}$$

$$|x_n| < \varepsilon + |\ell|, \quad \forall n > N. \tag{**}$$

If we now call $M = \max\{|x_1|, |x_2|, |x_3|, |x_4|, ..., |x_N|, \varepsilon + |\ell|\}$, then obviously

$$|x_n| < M \iff -M < x_n < M, \quad \forall n = 1,2,3, ... \tag{***}$$

showing thus that the sequence (x_n) is bounded, (see Definition 2-3 and Theorem 2-1)).

Example 9-3.

Show equation (9-5).

Solution

$$|x_n \cdot y_n - \ell \cdot m| = |x_n(y_n - m) + m(x_n - \ell)| \le |x_n||y_n - m| + |m||x_n - \ell|,$$

or since the sequence (x_n) is absolutely bounded, (there exists a constant M such that $|x_n| \le M$, $\forall n = 1,2,3, ...$, (see Example 9-2)) and $|m| < |m| + 1$,

$$|x_n \cdot y_n - \ell \cdot m| < M \cdot |x_n - m| + (|m| + 1) \cdot |x_n - \ell|. \tag{*}$$

Since $\lim y_n = m$, $(y_n - m)$ is a null sequence, and therefore $(|y_n - m|)$ and $(M \cdot |y_n - m|)$ will be null sequences, as well, (see (8-4) and (8-3), respectively). This means that given any $\varepsilon > 0$, there exists a positive integer N_1, such that

$$M \cdot |y_n - m| < \frac{\varepsilon}{2}, \qquad \forall\, n > N_1, \qquad\qquad\qquad (**)$$

and reasoning similarly,

$$(|m| + 1) \cdot |x_n - \ell| < \frac{\varepsilon}{2}, \qquad \forall\, n > N_2. \qquad\qquad (***)$$

If $N = max\,(N_1, N_2)$, then $\forall\, n > N$ inequalities $(**)$ and $(***)$ are **satisfied simultaneously**, and adding together yields,

$$M \cdot |y_n - m| + (|m| + 1) \cdot |x_n - \ell| < \varepsilon, \qquad \forall\, n > N,$$

or taking $(*)$ into consideration,

$$|x_n \cdot y_n - \ell \cdot m| < \varepsilon, \qquad \forall\, n > N, \qquad\qquad\qquad (****)$$

meaning that

$$\lim(x_n \cdot y_n) = \ell \cdot m = (\lim x_n) \cdot (\lim y_n),$$

and the proof is completed.

Example 9-4.
If a sequence (x_n) tends to **a positive limit** ℓ, then from a certain stage on, the terms of the sequence exceed the number $(\frac{\ell}{2})$. In symbols,

If $\lim x_n = \ell > 0$, then $\exists\, N : x_n > \frac{\ell}{2}$, $\forall\, n > N$.

Solution

Since, by assumption $\lim x_n = \ell > 0$,

$$\forall \varepsilon > 0 \quad \exists N = N(\varepsilon) : |x_n - \ell| < \varepsilon, \quad \forall n > N. \qquad\qquad (*)$$

Choosing $\varepsilon = \frac{\ell}{2}$, inequality $(*)$ yields,

$\exists N = N\left(\frac{\ell}{2}\right) : |x_n - \ell| < \frac{\ell}{2}, \ \forall n > N,$ or

$-\frac{\ell}{2} < x_n - \ell < \frac{\ell}{2} \iff \frac{\ell}{2} < x_n < \frac{3\ell}{2}, \quad \forall n > N,$

showing thus that from a certain stage on, i.e. $(\forall \, n > N = N\left(\frac{\ell}{2}\right))$, $x_n > \frac{\ell}{2}$,

$(x_{N+1} > \frac{\ell}{2}, \ x_{N+2} > \frac{\ell}{2}, \ x_{N+3} > \frac{\ell}{2}, x_{N+4} > \frac{\ell}{2}, \cdots)$, and the proof is completed.

Example 9-5.

Show that if $\lim x_n = \ell \neq 0$, then $\lim\left(\frac{1}{x_n}\right) = \frac{1}{\ell} = \frac{1}{\lim x_n}$.

Solution

a) Let $\ell > 0$. Then,

$\left|\frac{1}{x_n} - \frac{1}{\ell}\right| = \frac{|x_n - \ell|}{|x_n| \cdot \ell}$ since $|\ell| = \ell$. $\qquad\qquad$ (*)

By virtue of example 9-4,

$\exists N_1 : \forall n > N_1 \Rightarrow |x_n| = x_n > \frac{\ell}{2} > 0 \iff$

$0 < \frac{1}{|x_n|} < \frac{2}{\ell} \iff 0 < \frac{2}{x_n \cdot \ell} < \frac{2}{\ell^2}.$ $\qquad\qquad$ (**)

Also, since $\lim x_n = \ell$, the sequence $(x_n - \ell)$ is null, meaning that, given an arbitrary $\varepsilon > 0$,

$\exists \, N_2 = N_2(\varepsilon) : \quad \forall n > N_2 \Rightarrow |x_n - \ell| < \frac{\ell^2}{2} \cdot \varepsilon.$ $\qquad$ (***)

Setting $N = \max(N_1, N_2)$, then $\forall n > N$, inequalities (**) and (***) are **satisfied simultaneously**, and multiplying together yields,

$$\frac{|x_n - \ell|}{|x_n|\ell} < \varepsilon, \quad \forall n > N,$$

or taking (*) into account,

$$\left|\frac{1}{x_n} - \frac{1}{\ell}\right| < \varepsilon, \qquad \forall n > N,$$

showing thus that $\lim\left(\frac{1}{x_n}\right) = \frac{1}{\ell} = \frac{1}{\lim x_n}$, and the proof is completed.

b) Let $\ell < 0$.

Then $\lim(-x_n) = -\ell > 0 \Rightarrow \lim(-\frac{1}{x_n}) = -\frac{1}{\ell} \Rightarrow \lim(\frac{1}{x_n}) = \frac{1}{\ell}$, and this completes the proof.

Example 9-6.

If $x_n = \left(\frac{3n+5}{7n+10}\right)^2$, find the $\lim x_n$.

Solution

$$\lim x_n = \lim \left(\frac{3n+5}{7n+10}\right)^2 = \left\{\lim \frac{3n+5}{7n+10}\right\}^2 = \left\{\lim \frac{3+\frac{5}{n}}{7+\frac{10}{n}}\right\}^2 = \left\{\frac{\lim\left(3+\frac{5}{n}\right)}{\lim\left(7+\frac{10}{n}\right)}\right\}^2 = \left(\frac{3+0}{7+0}\right)^2 = \frac{9}{49}.$$

Example 9-7.

If $x_n = \frac{9n^5 - 13n^2 + 5}{18n^5 + 7n^2 - 1}$ find the $\lim x_n$.

Solution

$$x_n = \frac{9n^5 - 13n^2 + 5}{18n^5 + 7n^2 - 1} = \frac{(9n^5 - 13n^2 + 5) \div n^5}{(18n^5 + 7n^2 - 1) \div n^5} = \frac{9 - \frac{13}{n^3} + \frac{5}{n^5}}{18 + \frac{7}{n^3} - \frac{1}{n^5}}$$

and taking the limits of both sides, we have,

$$\lim x_n = \lim \frac{9 - \frac{13}{n^3} + \frac{5}{n^5}}{18 + \frac{7}{n^3} - \frac{1}{n^5}} = \frac{\lim(9 - \frac{13}{n^3} + \frac{5}{n^5})}{\lim(18 + \frac{7}{n^3} - \frac{1}{n^5})} = \frac{9 - 0 + 0}{18 + 0 - 0} = \frac{9}{18} = \frac{1}{2}.$$

Note: Example 9-7 is a characteristic Example. Let us suppose that **a sequence** (x_n) **is expressed as the quotient of two polynomials in** n, i.e.

$$x_n = \frac{a_k n^k + a_{k-1} n^{k-1} + \cdots + a_1 n + a_0}{b_m n^m + b_{m-1} n^{m-1} + \cdots + b_1 n + b_0}. \qquad (*)$$

In order to find the $\lim x_n$ as $n \to \infty$, **we divide both numerator and denominator by the highest power of** n, **appearing in** (*).

There are three cases:

a) If $k = m$, then $\lim x_n = \frac{a_k}{b_k}$, $(b_m = b_k)$.

b) If $k < m$, then $\lim x_n = 0$, while

c) If $k > m$, then $\lim x_n = +\infty$, if $a_k b_m > 0$,

$$\text{or } \lim x_n = -\infty, \text{ if } a_k b_m < 0.$$

Having applied this general rule in Example 9-7, we should have obtained immediately, $\lim x_n = \frac{9}{18} = \frac{1}{2}$, (case (a)).

Example 9-8.

If $x_n = \frac{3n}{7n+5}$, $y_n = 1 - \frac{1}{3^n}$, $w_n = \left(2 + \frac{1}{n}\right)^3$ find the $\lim(x_n \cdot y_n \cdot w_n)$.

Solution

$\lim x_n = \frac{3}{7}$, (see Note (a) in previous Example),

$\lim y_n = \lim(1 - \frac{1}{3^n}) = 1 - 0 = 1$, (see (8-13), with $x = \frac{1}{3}$), and

$\lim w_n = \lim\left(2 + \frac{1}{n}\right)^3 = \left\{\lim\left(2 + \frac{1}{n}\right)\right\}^3 = \{2 + 0\}^3 = 8$, (see (9-6)), therefore,

$\lim(x_n \cdot y_n \cdot w_n) = \lim x_n \cdot \lim y_n \cdot \lim w_n = \frac{3}{7} \cdot 1 \cdot 8 = \frac{24}{7}$.

Example 9-9.

If $x_n = \frac{n^3+7n^2-8}{n^4+10}$, find the $\lim x_n$.

Solution

Making use of Note (b), in Example 9-7, the $\lim x_n = 0$, (since 3=degree of numerator is less than 4 = degree of denominator).Let the reader go through the problem, step by step, dividing both terms ,numerator and denominator, by the highest power of n appearing in x_n, i.e. by n^4,etc.

Example 9-10.

If $x_n = \frac{\sqrt{5n^2+10n+1}}{3n+7}$, find the $\lim x_n$.

Solution

$$x_n = \frac{\sqrt{5n^2+10n+1}}{3n+7} = \frac{n\sqrt{5+\frac{10}{n}+\frac{1}{n^2}}}{n\left(3+\frac{7}{n}\right)} = \frac{\sqrt{5+\frac{10}{n}+\frac{1}{n^2}}}{3+\frac{7}{n}} \quad \text{therefore,}$$

$$\lim x_n = \lim \frac{\sqrt{5+\frac{10}{n}+\frac{1}{n^2}}}{3+\frac{7}{n}} = \frac{\sqrt{\lim\left(5+\frac{10}{n}+\frac{1}{n^2}\right)}}{\lim\left(3+\frac{7}{n}\right)} = \frac{\sqrt{5+0+0}}{3+0} = \frac{\sqrt{5}}{3}.$$

Example 9-11.

Show that the sequence with general term $x_n = \frac{3n-1}{5n+2}$ is increasing and bounded, and find its limit.

Solution

$$x_{n+1} > x_n \Longleftrightarrow \frac{3(n+1)-1}{5(n+1)+2} > \frac{3n-1}{5n+2} \Longleftrightarrow \frac{3n+2}{5n+7} > \frac{3n-1}{5n+2} \Longleftrightarrow (3n+2)(5n+2) >$$
$$(5n+7)(3n-1) \Longleftrightarrow 15n^2 + 16n + 4 > 15n^2 + 16n - 7,$$

or, $4 > -7$, which is true for all $n \in \mathbb{N}$, and going backwards, one easily obtains that $x_{n+1} > x_n$, for $n = 1,2,3,\cdots$, i.e. the sequence (x_n) is increasing. Also, the given sequence is bounded, the reader may easily prove that $0 < x_n < 1, \forall n \in \mathbb{N}$. Finally, the limit of (x_n) does exist, and is $\lim x_n = \frac{3}{5}$, according to note (a), in Example 9-2.

Example 9-12.

If $x_n = \sqrt[3]{n+5} - \sqrt[3]{n+2}$, find the $\lim x_n$.

Solution

Applying the well known, elementary identity,

$$a^3 - b^3 = (a-b)(a^2 + ab + b^2) \Longleftrightarrow a - b = \frac{a^3-b^3}{a^2+ab+b^2},$$

for $a = \sqrt[3]{n+5}$ and $b = \sqrt[3]{n+2}$, one obtains

$$x_n = \sqrt[3]{n+5} - \sqrt[3]{n+2} = \frac{(n+5)-(n+2)}{\left(\sqrt[3]{n+5}\right)^2 + \left(\sqrt[3]{n+5}\right)\left(\sqrt[3]{n+2}\right) + \left(\sqrt[3]{n+2}\right)^2} =$$

$$\frac{3}{(n+5)^{2/3}+(n+5)^{1/3}(n+2)^{1/3}+(n+2)^{2/3}}.$$

As $n \to \infty$, the denominator tends to $+\infty$, (see (8-7), (8-8), (8-9) and (8-10)), while the numerator remains constant, therefore, (see (9-11)),

$$\lim x_n = \lim\left(\sqrt[3]{n+5} - \sqrt[3]{n+2}\right) = 0.$$

Example 9-13.

Show that the sequence $y_n = \dfrac{3 \cdot 2^n - n}{5 \cdot 2^n + n}$ converges to the number $\dfrac{3}{5}$.

Solution

The $n^{\underline{th}}$ term of the sequence $y_n = \dfrac{3 \cdot 2^n - n}{5 \cdot 2^n + n} = \dfrac{3 - \frac{n}{2^n}}{5 + \frac{n}{2^n}}$. $\qquad$ (*)

As $n \to \infty$, the sequence $\dfrac{n}{2^n} \to 0$, (see (8-15)), therefore,

$$\lim\left(3 - \frac{n}{2^n}\right) = 3 - 0 = 3 \text{ and } \lim\left(5 + \frac{n}{2^n}\right) = 5 + 0 = 5, \text{ from which,}$$

$$\lim y_n = \lim \frac{3 - \frac{n}{2^n}}{5 + \frac{n}{2^n}} = \frac{\lim\left(3 - \frac{n}{2^n}\right)}{\lim\left(5 + \frac{n}{2^n}\right)} = \frac{3 - 0}{5 + 0} = \frac{3}{5}.$$

Example 9-14.

If $x_n = \dfrac{1 + 2 + 3 + \cdots + n}{n^2}$, find the $\lim x_n$.

Solution

It is well known that $1 + 2 + 3 + \cdots + n = \dfrac{n(n+1)}{2}$, therefore $x_n = \dfrac{n(n+1)}{2n^2}$, and

$$\lim x_n = \lim \frac{n^2 + n}{2n^2} = \frac{1}{2}.$$

(see Note (a), in Example 9-7).

Example 9-15.

If $y_n = \sqrt[3]{\dfrac{(5 + \sqrt{n})(7 - \sqrt{n})}{27n + 15}}$, find the $\lim y_n$.

Solution

The under the cubic root quantity, is

$$b_n = \frac{-n+2\sqrt{n}+35}{27n+15} = \frac{(-n+2\sqrt{n}+35)\div n}{(27n+15)\div n} = \frac{-1+\frac{2}{\sqrt{n}}+\frac{35}{n}}{27+\frac{15}{n}},$$

and as $n \to \infty$, $b_n \to -\frac{1}{27}$, therefore,

$$\lim y_n = \lim \sqrt[3]{b_n} = \sqrt[3]{\lim b_n} = \sqrt[3]{\left(-\frac{1}{27}\right)} = -\frac{1}{3}.$$

PROBLEMS

9-1) Using Example 9-5, and property (9-5), show property (9-7).

Hint: $\left(\frac{x_n}{y_n}\right) = x_n \cdot \left(\frac{1}{y_n}\right)$.

9-2) Show property (9-8) .

Hint: Use the well known identity,

$$a^k - b^k = (a - b)(a^{k-1} + a^{k-1} \cdot b + \cdots + a \cdot b^{k-2} + b^{k-1}),$$

apply it for $a = \sqrt[k]{x_n}$ and $b = \sqrt[k]{\ell}$, and take into consideration Example 9-4.

9-3) Making use of properties (9-6) and (9-8), show property (9-9).

9-4) If $x_n = \frac{n+1}{n}$, $y_n = 1 - \frac{1}{4^n}$ and $z_n = \frac{n^2+5}{3n^2-7}$, find the $\lim(x_n \cdot y_n \cdot z_n)$.

9-5) If $x_n = \frac{2n+3}{3n+1}$ and $y_n = 4 + \frac{1}{5^n}$ find,

a) $\lim(x_n \cdot y_n)$ and **b)** $\lim\left(\frac{x_n}{y_n}\right)$.

(**Answer: a)** $\frac{8}{3}$, **b)** $\frac{1}{6}$).

9-6) If $x_n = \frac{3n^7+5}{5n^7+16n^2-1}$ and $y_n = \frac{2n^2+1}{n^2-7}$ find,

a) $\lim x_n$, **b)** $\lim(x_n)^2$, **c)** $\lim y_n$, **d)** $\lim(y_n)^5$,**e)** $\lim\left(\sqrt[3]{x_n} \cdot (y_n)^2\right)$.

9-7) If $x_n = \frac{-n^3+1}{8n^3+7}$, find

a) $\lim x_n$, **b)** $\lim(x_n)^2$, **c)** $\lim \sqrt[3]{x_n}$.

(**Answer: a)** $-\frac{1}{8}$,   **b)** $\frac{1}{64}$,   **c)** $-\frac{1}{2}$).

9-8) If $x_n = \frac{1^2+2^2+\cdots+n^2}{n^3}$, find the $\lim x_n$.

Hint: $1^2 + 2^2 + \cdots + n^2 = \frac{1}{6}n(n+1)(2n+1)$, (see Problem 9-24).

9-9) If $y_n = \frac{1^3+2^3+\cdots+n^3}{n^4}$, find the $\lim y_n$.

Hint: $1^3 + 2^3 + \cdots + n^3 = \left\{\frac{n(n+1)}{2}\right\}^2$, (see Problem 9-24).

(**Answer:** $\frac{1}{4}$).

9-10) If (x_n) and (y_n) are the sequences given in Problems 9-8 and 9-9, respectively, show that $\lim \sqrt[3]{x_n y_n} = \frac{1}{\sqrt[3]{12}}$ and that $\lim \sqrt[3]{\frac{x_n}{y_n}} = \sqrt[3]{\frac{4}{3}}$.

9-11) If (x_n) is a bounded sequence and (y_n) is a null sequence, show that the sequence $(x_n y_n)$ is null.

As an application, show that $\lim \frac{\sin(n^2+5)}{n} = 0$, and $\lim \frac{\cos(n)}{n^3+10} = 0$.

9-12) If $x_n = \frac{-n^3+5}{n^2+1}$ and $y_n = \frac{7n^2+10}{n^3+2}$, show that

$\lim x_n = -\infty$, $\lim y_n = 0$, and $\lim(x_n y_n) = -7$.

9-13) If c is any fixed number, find the limit of the sequence,

$$x_n = \frac{1}{n}\left\{\left(c+\frac{1}{n}\right)^2 + \left(c+\frac{2}{n}\right)^2 + \cdots + \left(c+\frac{n-2}{n}\right)^2 + \left(c+\frac{n-1}{n}\right)^2\right\}.$$

(Answer: $c^2 + c + \frac{1}{3}$).

9-14) If $x_n = \sqrt{(n+1)(n+5)} - n$, show that $\lim x_n = 3$.

9-15) If $x_n = \sqrt[3]{\frac{(2+\sqrt{n})(3+\sqrt{n})}{64n+1}}$, $y_n = \frac{n}{\sqrt[3]{27n^3+n}-n}$, find,

a) $\lim x_n$, **b)** $\lim y_n$, **c)** $\lim(x_n + y_n)$, **d)** $\lim(x_n y_n)$.

(Answer: **a)** $\frac{1}{4}$ **b)** $\frac{1}{2}$ **c)** $\frac{3}{4}$ **d)** $\frac{1}{8}$).

9-16) If $x_n = \left(\frac{n+1}{2n+3}\right)^5$ and $y_n = \frac{2+\frac{n}{2^n}}{3+\frac{7n}{3^n}}$, show that

$\lim x_n = \frac{1}{32}$, $\lim y_n = \frac{2}{3}$, and that the sequences $(x_n y_n)$ and $\left(\frac{x_n}{y_n}\right)$ are bounded.

Hint: See Example 9-2.

9-17) If $x_n = \frac{(n+2\cdot 7^n)\cdot 5^n}{(-n+16\cdot 5^n)\cdot 7^n}$, find $\lim x_n$, $\lim(x_n)^2$, and $\lim(\sqrt[3]{x_n})$.

(Answer: $\frac{1}{8}, \frac{1}{64}, \frac{1}{2}$).

9-18) Show that $\lim \frac{(-3)^n - 2n}{(-3)^n + 3n} = 1$.

9-19) If $x_n = n^3\left\{\sqrt{n^2 + \sqrt{n^4+1}} - \sqrt{2}n\right\}$, find $\lim x_n$.

Hint: $x_n = n^3 \frac{\left(\sqrt{n^2+\sqrt{n^4+1}}-\sqrt{2}n\right)\left(\sqrt{n^2+\sqrt{n^4+1}}+\sqrt{2}n\right)}{\sqrt{n^2+\sqrt{n^4+1}}+\sqrt{2}n} = n^3 \frac{n^2+\sqrt{n^4+1}-2n^2}{\sqrt{n^2+\sqrt{n^4+1}}+\sqrt{2}n} =$

$n^3 \frac{\sqrt{n^4+1}-n^2}{\sqrt{n^2+\sqrt{n^4+1}}+\sqrt{2}n} = n^3 \frac{\left(\sqrt{n^4+1}-n^2\right)\left(\sqrt{n^4+1}+n^2\right)}{\left(\sqrt{n^2+\sqrt{n^4+1}}+\sqrt{2}n\right)\left(\sqrt{n^4+1}+n^2\right)} =$

$n^3 \frac{(n^4-1)-n^4}{\left(\sqrt{n^2+\sqrt{n^4+1}}+\sqrt{2}n\right)\left(\sqrt{n^4+1}+n^2\right)} = \frac{n^3}{\left(\sqrt{n^2+\sqrt{n^4+1}}+\sqrt{2}n\right)\left(\sqrt{n^4+1}+n^2\right)}$, etc.

(Answer: $\frac{1}{4\sqrt{2}}$).

9-20) If c and d, are positive constants, and

$$x_n = \sqrt{n^2 + 2 \cdot d \cdot n + 1} - \sqrt{n^2 + c},$$

show that $\lim x_n = d$.

Hint: $x_n = \frac{(\sqrt{n^2+2\cdot d\cdot n+1}-\sqrt{n^2+c})(\sqrt{n^2+2\cdot d\cdot n+1}+\sqrt{n^2+c})}{\sqrt{n^2+2\cdot d\cdot n+1}+\sqrt{n^2+c}} = \frac{(n^2+2\cdot d\cdot n+1)-(n^2+c)}{\sqrt{n^2+2\cdot d\cdot n+1}+\sqrt{n^2+c}}$, etc.

9-21) If $x_n = \left(\sqrt{n+5} - \sqrt{n}\right)\sqrt{n+3}$, find $\lim x_n$.

(Answer: $\frac{5}{2}$).

9-22) If $y_n = \frac{n^3}{5n^2+7} - \frac{n^2}{5n+7}$, find $\lim y_n$.

9-23) For the sequence (x_n), in Problem 9-21, find **a)** $\lim(x_n)^{2/3}$, **b)** $\lim(x_n)^{1/7}$, and **c)** $\lim \sqrt[5]{(x_n)^4}$.

(Answer: **a)** $\left(\frac{5}{2}\right)^{\frac{2}{3}}$, **b)** $\sqrt[7]{\frac{5}{2}}$, **c)** $\left(\frac{5}{2}\right)^{\frac{4}{5}}$).

9-24) If n is any positive integer, show that,

a) $1 + 2 + 3 + \cdots + n = \frac{n(n+1)}{2}$,

b) $1^2 + 2^2 + 3^2 + \cdots + n^2 = \frac{1}{6}n(n+1)(2n+1)$,

c) $1^3 + 2^3 + 3^3 + \cdots + n^3 = \left(\frac{n(n+1)}{2}\right)^2$.

9-25) If $x_n = \frac{\sin n}{\sqrt[3]{n}}$ and $y_n = \frac{\cos \sqrt{n}}{n+3}$, show that $\lim\left(x_n^2 \cdot \sqrt[3]{y_n}\right) = 0$.

9-26) The sequence $(a_n + b_n)$ is convergent to a finite number $\ell \in \mathbb{R}$. Can we assert that the sequences (a_n) and (b_n) are necessarily convergent?

9-27) Find the limit of the following sequences,

a) $x_n = \dfrac{(2+3n^4)(1-2n-n^3)}{(n^3+7)(4+n^4)}$ **b)** $y_n = \left(8 + \dfrac{1}{n^3}\right)\left(\dfrac{n^2-7}{n^2+6}\right)$.

(**Answer: a)** -3 , **b)** 8).

9-28) The sequences (x_n) and (y_n) are divergent. Can we assert that the sequence $(x_n + y_n)$ is necessarily divergent?

9-29) Find the limit of the following sequences,

a) $x_n = \dfrac{7}{n^3}(1^2 + 2^2 + 3^2 + \cdots + n^2)$ **b)** $y_n = \dfrac{(n+1)!+(n+2)!}{(n+3)!}$

(**Answer: a)** $\dfrac{14}{6}$ **b)** 0).

9-30) Find the limit of the following sequences,

a) $x_n = \dfrac{(2n+1)^3}{5n^3-7}$ **b)** $y_n = \dfrac{c^n-c^{-n}}{c^n+c^{-n}}$ where c is a fixed real number $\neq 0$.

10. General Theorems on Limits.

Given a sequence (x_n), there are two basic, fundamental questions, to be answered:

1) Does (x_n) converges to a finite limit, or it diverges to infinity, and

2) If (x_n) converges, what is its limit?

Answers to both questions are achieved by means of some general Theorems and techniques, which are to be developed in the sequel of this chapter.

We should point out that the basic definition for the limit of a sequence,(eq.(6-1)),is rarely used in practice, in order to find the limit of the sequence, for the simple reason, that in (6-1) the limit ℓ is supposed to be known in advance,something which obviously is not the case. On the contrary, the limit ℓ is not known, and the main problem, in sequences, is to determine the limit ℓ, given the sequence (x_n).

The following Theorems offer a great aid, towards this direction.

Theorem 10-1. (Uniqueness of the limit).
The limit of a convergent sequence (x_n) is unique, or equivalently,

If $\lim x_n = \ell_1$ and $\lim x_n = \ell_2$, then $\ell_1 = \ell_2$.

Because of this Theorem, we may speak about **the limit** of a convergent sequence.

Proof: Assuming that (x_n) has two limits, ℓ_1 and ℓ_2, we shall show that $\ell_1 = \ell_2$.

From $\lim x_n = \ell_1$, we have that,

$$\forall \varepsilon > 0 \quad \exists N_1 = N_1(\varepsilon): \quad |x_n - \ell_1| < \frac{\varepsilon}{2} \quad \forall n > N_1. \qquad (*)$$

Similarly, from $\lim x_n = \ell_2$, we have,

$$\forall \varepsilon > 0 \quad \exists N_2 = N_2(\varepsilon): \quad |x_n - \ell_2| < \frac{\varepsilon}{2} \quad \forall n > N_2. \qquad (**)$$

Let $N = max(N_1, N_2)$. Then for all $\boldsymbol{n} > N$,

$$|x_n - \ell_1| < \frac{\varepsilon}{2} \text{ and } |x_n - \ell_2| < \frac{\varepsilon}{2} \quad \forall n > N, \qquad (***)$$

and adding together yields, $\quad |x_n - \ell_1| + |x_n - \ell_2| < \varepsilon, \quad \forall n > N. \quad (****)$

However,

$$0 \le |\ell_2 - \ell_1| = |(x_n - \ell_1) - (x_n - \ell_2)| \le |x_n - \ell_1| + |x_n - \ell_2| < \varepsilon, \quad \forall n > N,$$

(from (****)), and **since ε is an arbitrarily small positive number**, necessarily $|\ell_1 - \ell_2| = 0$, or $\ell_1 = \ell_2$, and the proof is completed.

Remark: At this point we should note that a given sequence could be neither convergent (to a finite limit) nor divergent to $+\infty$ or to $-\infty$. For example, the sequence $x_n = (-1)^n n$, i.e. $\{x_1 = -1, x_2 = 2, x_3 = -3, x_4 = 4, x_5 = -5, \cdots\}$, belongs to this case. Another example could be the sequence $y_n = \cos(n\pi) n^2$, etc.

Theorem 10-1, simply states that **if a given sequence (x_n) converges to a finite limit, then this limit is unique.**

Theorem 10-2.
Every convergent sequence (x_n) is bounded.

Proof: For a proof, see Example 9-2.

Theorem 10-3.
If a sequence (x_n) tends to a positive limit ℓ, then from a certain stage on, the terms of the sequence exceed the number $\frac{\ell}{2}$. In symbols,

If $\lim x_n = \ell > 0$, then $\exists N: x_n > \frac{\ell}{2}, \quad \forall n > N$.

Proof: For a proof, see Example 9-4.

Theorem 10-4. (Trapped sequences).
If two sequences (a_n) and (b_n) converge to the same limit ℓ, and if (x_n) is

another sequence such that $a_n \le x_n \le b_n$, $\forall n \ge 1$, i.e. if (x_n) is trapped between (a_n) and (b_n), then the sequence x_n also converges to the same limit ℓ. In symbols,

If $\lim a_n = \lim b_n = \ell$, and if $a_n \le x_n \le b_n$, $\forall n \ge 1$, then $\lim x_n = \ell$.

Proof: Let $\varepsilon > 0$ be a given arbitrarily small positive number. There exist two integers N_1 and N_2, such that $|a_n - \ell| < \varepsilon$, $\forall n > N_1$ and $|b_n - \ell| < \varepsilon$, $\forall n > N_2$.　　　　　　　　　　　　(*)

If $N = \max(N_1, N_2)$, then both inequalities in (*) hold for all $n > N$. In particular,

$\ell - \varepsilon < a_n$ and $b_n < l + \varepsilon$, and since $a_n \le x_n \le b_n$, we have,

$\ell - \varepsilon < a_n \le x_n \le b_n < l + \varepsilon \Rightarrow \ell - \varepsilon < x_n < l + \varepsilon \Leftrightarrow |x_n - \ell| < \varepsilon$, $\forall n > N$,

meaning that $\lim x_n = \ell$, and this completes the proof.

Corollary 10-1: If $0 < x_n < y_n$, and $\lim y_n = 0$, then $\lim x_n = 0$, as well.

As an application, one may show easily that the sequences $x_n = \frac{(\sin(n+5))^2}{n}$, and $y_n = \frac{(\cos n)^4}{\sqrt{n}}$ are null sequences.

Remarks: a) Theorem 10-4, is still valid in cases where the sequence (x_n) is eventually trapped between the two sequences (a_n) and (b_n), i.e. if $a_n \le x_n \le b_n$, $\forall n \ge k$, where k is a fixed positive integer.

b) If $a_n \le x_n$, $\forall n \ge k$ and $\lim a_n = +\infty$, then $\lim x_n = +\infty$, as well. The proof is easy, let the reader try it.

Next Theorem is a very useful and powerful Theorem, used widely in practice, in order to show the convergence of a given sequence.

Theorem 10-5. (Convergence of monotone and bounded sequences).
a) If (x_n) is an increasing sequence, bounded above by a number M, then (x_n) is convergent to a limit $\le M$.

b) If (y_n) is a decreasing sequence, bounded below by a number m, then (y_n) is convergent to a limit $\geq m$. In symbols,

If $x_1 < x_2 < x_3 < \cdots < x_n < x_{n+1} < \cdots \leq M$, **then** $\lim x_n \leq M$.

If $y_1 > y_2 > y_3 > \cdots > y_n > y_{n+1} > \cdots \geq m$, **then** $\lim y_n \geq m$.

Note: a) If there is no upper bound for an increasing sequence (x_n), then $\lim x_n = +\infty$.

b) If there is no lower bound for a decreasing sequence (y_n), then $\lim y_n = -\infty$.

In order to prove Theorem 10-5, we will make use of the following axiom, about real numbers, known as the Completeness Axiom.

The Completeness Axiom:

If a nonempty set X of **real numbers** has an upper bound, then X has a smallest upper bound, called **the least upper bound (l.u.b) or supremum of X, (supX).**

Based on this axiom, one may show that, if a nonempty set Y of real numbers has a lower bound, then Y has a largest lower bound, called **the greatest lower bound (g.l.b) or infimum of Y, (infY).**

(It suffices to consider the set having elements the opposite of the elements of Y).

Obviously, if a set has upper and lower bounds, then this set has a (sup) and an (inf). As an example, let us consider the set A of real numbers in the interval $(7, \cdots, 10)$, i.e. $A = \{x \in \mathbb{R} : 7 < x < 10\}$. Any number greater than 10, for example $11, 12.1, 12.8, 15$, etc is an upper bound of A. The set A has **an infinite number of upper bounds**. However the smallest upper bound of A, is the number 10, which is actually the (sup) of A. Similarly, the number 7, is the (inf) of A.

Proof of Theorem 10-5:

a) Let us assume that M is an upper bound of x_n, i.e. $x_n \leq M, \ \forall n$. Let $X = \{x_1, x_2, x_3, \cdots, x_n, x_{n+1}, \cdots\}$. According to the completeness axiom the set X will have a (l.u.b) or supremum, call it ℓ, i.e. $\boldsymbol{\ell} =\textbf{sup}(\boldsymbol{X})\leq \boldsymbol{M}$. Now let $\varepsilon > 0$ be any given, arbitrarily small positive number. Since ℓ is the $\sup(X)$, the number $(\ell - \varepsilon)$ is not an upper bound of X, meaning that there exists **at least one term** x_N, such that $x_N > (l - \varepsilon)$, and since (x_n) is increasing, we have,

$$x_n > x_N > l - \varepsilon, \quad \forall n > N.$$

At the same time, (since no terms of (x_n) can exceed ℓ),

$$\ell + \varepsilon > l > x_n > l - \varepsilon \Leftrightarrow -\varepsilon < x_n - \ell < \varepsilon \Leftrightarrow |x_n - \ell| < \varepsilon, \ \forall n > N.$$

This last inequality shows that $\lim x_n = \ell \leq M$, and this completes the proof.

b) Quite similarly, one may show part (b) of the Theorem.

Theorem 10-6. (Principle of nested intervals).
Let (x_n) and (y_n) be two given sequences. If
a) $(x_n) \nearrow$ ((x_n) is increasing or non-decreasing) and $(y_n) \searrow$ ((y_n) is decreasing
or   non-increasing),

b) $\forall n \in \mathbb{N}, \ x_n < y_n$, and

c) $\lim(y_n - x_n) = 0$,

then both sequences are convergent to a common limit ℓ, i.e.
$\lim x_n = \lim y_n = \ell$.

Proof: Since $(x_n) \nearrow, (y_n) \searrow$ and $x_n < y_n, \ \forall n \in \mathbb{N}$, we have,

$$x_1 \leq x_2 \leq x_3 \leq \cdots \leq x_n \leq x_{n+1} \leq \cdots \leq y_{n+1} \leq y_n \leq \cdots \leq y_3 \leq y_2 \leq y_1.$$

The increasing sequence (x_n) has an upper bound,(for example the term y_1),therefore, according to Theorem 10-5,is convergent ,and let $\lim x_n = \ell_1$.

Similarly, the decreasing sequence (y_n) has a lower bound,(for example the term x_1),therefore again by virtue of Theorem 10-5,is convergent, and let $\lim y_n = \ell_2$.

It remains to show that $\ell_1 = \ell_2$.

The $\lim(y_n - x_n) = \lim y_n - \lim x_n = \ell_2 - \ell_1 = 0$, because of the assumption (c),therefore $\ell_1 = \ell_2$, and the proof is completed.

Note: Theorem 10-6, may be expressed graphically, with the aid of Fig.10-1.

Let $D_1 = [x_1, y_1], D_2 = [x_2, y_2], D_3 = [x_3, y_3], \cdots D_n = [x_n, y_n], \cdots$ be a sequence of intervals, such that $D_1 \supset D_2 \supset D_3 \supset \cdots \supset D_n \supset \cdots$.Then if $\lim(y_n - x_n) = 0$, there exist **one and only one point ℓ common to all the intervals.**

In this case, the intervals $D_i, \quad i = 1,2,3,\cdots$ are called **nested intervals**, and the corresponding principle, is known as **the nested intervals principle**.

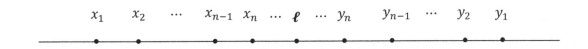

Fig.10-1: The nested intervals principle.

Theorem 10-7.

Let (x_n) be a convergent sequence, tending to a limit ℓ, as $n \to \infty$, i.e.$\lim x_n = \ell$. Then

$$\lim x_n = \lim x_{n+1} = \lim x_{n+2} = \cdots = \lim x_{n+k} = \ell,$$

where **k is a fixed positive integer**, not depending on n.

Proof: Obvious.

For example, if $(x_n) = \{x_1, x_2, x_3, x_4, \cdots\}$, then $(x_{n+1}) = \{x_2, x_3, x_4, x_5, \cdots\}$, $(x_{n+2}) = \{x_3, x_4, x_5, x_6 \cdots\}$, etc. Obviously, the sequences $(x_{n+1}), (x_{n+2}), \cdots$ tend to the same limit $\ell = \lim x_n$.

Theorem 10-7 is used widely in practice, in order to evaluate limits of sequences defined recursively, (see Equations (1-2) and (1-3)).

Theorem 10-8.

If a sequence (x_n) converges to a limit ℓ, then every subsequence of (x_n) converges to the same limit ℓ.

Proof: The fact that $\lim(x_n) = \ell$, means that from a certain stage on, all the terms of (x_n) fall within the interval $(\ell - \varepsilon, \ell + \varepsilon)$, where $\varepsilon > 0$ is a given, arbitrarily small positive number,(see Chapter 6).Since the terms of any subsequence of (x_n), are among the terms of (x_n), the terms of the subsequence eventually fall into the same interval $(\ell - \varepsilon, \ell + \varepsilon)$, which in turns means that the subsequence tends to the limit ℓ, and this completes the proof.

Corollary 10-2: If two subsequences of (x_n) tend to two different limits, then the sequence (x_n) has no limit.

Theorem 10-9. (The quotient rule).

Let (x_n) be a given sequence. If $\lim \left| \frac{x_{n+1}}{x_n} \right| = c$, where $0 \leq c < 1$, then the $\lim x_n = 0$, (i.e.(x_n) is a null sequence).

Proof: Since $\lim \left| \frac{x_{n+1}}{x_n} \right| = c$, where $0 \leq c < 1$,

$$\forall \varepsilon > 0, \ \exists N = N(\varepsilon): \ \forall n > N \Rightarrow c - \varepsilon < \left| \frac{x_{n+1}}{x_n} \right| < c + \varepsilon,$$

or equivalently, $|x_{n+1}| < (c + \varepsilon)|x_n|, \quad \forall n > N.$ $\qquad$ (*)

We may choose $\varepsilon > 0$, so small that $(c + \varepsilon) < 1$. (Note that $(c + \varepsilon) > 0$).

Applying (*) for $n = N + 1, N + 2, N + 3, \cdots N + k$, we have,

$$|x_{N+2}| < |x_{N+1}|(c + \varepsilon)$$

$$|x_{N+3}| < |x_{N+2}|(c + \varepsilon)$$

$$\cdots \cdots \cdots \cdots$$

$$|x_{N+k+1}| < |x_{N+k}|(c + \varepsilon)$$

and multiplying together, yields, $\quad 0 < |x_{N+k+1}| < |x_{N+1}|(c + \varepsilon)^k.$ (**)

As $k \to \infty, (c + \varepsilon)^k \to 0$, (see Theorem 8-1), while $|x_{N+1}|$ is just a constant number, therefore, $|x_{N+1}|(c + \varepsilon)^k \to 0$, as $k \to \infty$. Making use of Corollary 10-1, we conclude that the sequence $(x_{k+(N+1)})$, $k = 1,2,3, \cdots$ is null, therefore (x_k) $k = 1,2,3, \cdots$ is null, or the same, the sequence (x_n), $n = 1,2,3, \cdots$ is null, and this completes the proof.

Theorem 10-10. (Cauchy 's Theorem about the $\lim \sqrt[n]{x_n}$).

Let (x_n) be a sequence with positive terms, $(x_n > 0, \ \forall n \in \mathbb{N})$. If $\lim \frac{x_{n+1}}{x_n} = c$, where c is a finite, non negative number, $(c \geq 0)$ or $c = +\infty$, then,

$$\lim \sqrt[n]{x_n} = \lim \frac{x_{n+1}}{x_n} = c.$$

In other words, the sequence $(\sqrt[n]{x_n})$ tends where the sequence $(\frac{x_{n+1}}{x_n})$ tends to, as $n \to \infty$. The inverse is not necessarily true.

The proof is omitted. The interested reader may try to prove the Theorem, first for the case where c is a finite number, and then for the case where $c = +\infty$.

Theorem 10-10 is a very useful Theorem, especially in cases where the nth root of a sequence is involved, and the limit is to be determined.

Theorem 10-11. (The Arithmetic Mean Theorem).

If $\lim x_n = \ell$, then $\lim \frac{x_1+x_2+x_3+\cdots+x_n}{n} = \ell.$

(The quantity $\frac{x_1+x_2+x_3+\cdots+x_n}{n}$ is by definition, the **arithmetic mean** of the n numbers, $x_1, x_2, x_3, \cdots, x_n$).

Proof: a) Let us assume, at first, that $\ell = 0$, i.e. (x_n) is a null sequence. Let us also define a new sequence (y_n), as

$$y_n = \frac{x_1+x_2+x_3+\cdots+x_n}{n} \qquad n = 1,2,3, \cdots$$

We shall show that (y_n) is a null sequence, as well. It suffices to show that **for large values of the index n**, the corresponding terms $|y_n|$ become **arbitrarily small**, i.e. smaller than any arbitrary small positive number $\varepsilon > 0$.

Since ,by assumption, $\lim x_n = 0$, eventually all the terms of (x_n), will in absolute value be smaller than any $\varepsilon > 0$, i.e.

$$|x_n| < \varepsilon, \ \forall n > N = N(\varepsilon) \implies |x_{N+1}| < \varepsilon, |x_{N+2}| < \varepsilon, |x_{N+3}| < \varepsilon, \cdots. \qquad (*)$$

Let us now consider the terms of (y_n), for $n \geq N + 1$.

$$|y_n| = \frac{|x_1 + x_2 + \cdots + x_N + x_{N+1} + x_{N+2} + \cdots + x_n|}{n} \leq \frac{|x_1| + |x_2| + \cdots + |x_N| + |x_{N+1}| + |x_{N+2}| + \cdots + |x_n|}{n} <$$
$$\frac{|x_1| + |x_2| + \cdots + |x_N| + (n-N)\varepsilon}{n}$$

where in the last inequality, (*) has been taken into account, and finally,

$$|y_n| < \frac{|x_1| + |x_2| + \cdots + |x_N|}{n} + \frac{n-N}{n}\varepsilon. \qquad (**)$$

Assuming n quite large, $(n \to \infty)$, the term $\frac{|x_1| + |x_2| + \cdots + |x_N|}{n}$ becomes **arbitrarily small**, since the numerator remains constant, while the denominator can become **arbitrarily large**, meaning that for quite large values of n,

$$\frac{|x_1| + |x_2| + \cdots + |x_N|}{n} < \varepsilon_1$$

where ε_1 **is an arbitrarily small positive number**. So ,for quite large values of n,from (**) and (***),we have,

$$|y_n| < \varepsilon_1 + \frac{n-N}{n}\varepsilon < \varepsilon_1 + \varepsilon. \qquad (***)$$

Since ε_1 and ε are arbitrarily small positive numbers, **their sum $(\varepsilon_1 + \varepsilon)$, will be another arbitrarily small positive number**, showing thus, that $\lim y_n = 0$.

b) Let us now assume that $\ell \neq 0$.Then the sequence $(w_n) = (x_n - \ell)$, will be a null sequence, i.e. $\lim w_n = 0$.According to the part (a),

$$\lim \frac{w_1+w_2+\cdots+w_n}{n} = 0 \Longleftrightarrow$$

$$\lim \frac{(x_1-\ell)+(x_2-\ell)+\cdots+(x_n-\ell)}{n} = 0 \Longleftrightarrow \lim \frac{(x_1+x_2+\cdots+x_n)-n\ell}{n} = 0 \Longleftrightarrow$$

$$\lim \left\{\frac{x_1+x_2+\cdots+x_n}{n} - \ell\right\} = 0 \Longleftrightarrow \lim \frac{x_1+x_2+\cdots+x_n}{n} = \ell,$$

and this completes the proof.

Note: If $\ell = +\infty$, Theorem 10-11, is still valid. For a proof, see Problem 10-23.

Theorem 10-12. (The Three Means Theorem).
Let (x_n) be a sequence with positive terms converging to a limit $\ell > 0$. We define three new sequences,

a) $y_n = \dfrac{x_1+x_2+x_3+\cdots x_n}{n}$,   **(The Arithmetic Mean),**

b) $z_n = \sqrt[n]{x_1 x_2 x_3 \cdots x_n}$,   **(The Geometric Mean),**

c) $w_n = \dfrac{n}{\frac{1}{x_1}+\frac{1}{x_2}+\frac{1}{x_3}+\cdots+\frac{1}{x_n}}$,   **(The Harmonic Mean).**

Then, $\lim y_n = \lim z_n = \lim w_n = \ell = \lim x_n$.

Proof: a) It was proved in Theorem 10-11.

b) Let $a_n = x_1 x_2 x_3 \cdots x_n$. By virtue of Theorem 10-10,

$$\lim \sqrt[n]{a_n} = \lim \frac{a_{n+1}}{a_n} \Longrightarrow \lim \sqrt[n]{x_1 x_2 x_3 \cdots x_n} = \lim \frac{x_1 x_2 x_3 \cdots x_n x_{n+1}}{x_1 x_2 x_3 \cdots x_n} = \lim x_{n+1} \Longrightarrow$$
$$\lim z_n = \lim x_{n+1} = \lim x_n = \ell,$$

and this completes the proof of part (b).

c) Since $\lim x_n = \ell$, the $\lim \dfrac{1}{x_n} = \dfrac{1}{\ell}$ (see Example 9-5), so according to part (a),

$$\lim \frac{\frac{1}{x_1}+\frac{1}{x_2}+\cdots+\frac{1}{x_n}}{n} = \frac{1}{\ell} \Longrightarrow \lim \frac{1}{w_n} = \frac{1}{\ell} \Longrightarrow \lim w_n = \ell,$$

and the proof is completed.

For a different proof of this Theorem, see Problem 10-22.

Note: Theorem 10-12, remains valid even if $\ell = +\infty$.

Example 10-1.

If $x_n = \left(1 + \frac{1}{n}\right)^n$ and $y_n = \left(1 + \frac{1}{n+1}\right)^n$, show that

a) Both sequences are convergent, and

b) $\lim x_n = \lim y_n$.

Solution

a) The sequence (x_n) is increasing and bounded above, while (y_n) is decreasing and bounded below,(see Examples 3-2 and 3-3).By virtue of Theorem 10-5,both sequences are convergent ,and let

$$\lim x_n = \lim \left(1 + \frac{1}{n}\right)^n = \ell, \text{(a finite number)}.$$

b) For the given sequences,

i) $(x_n) \nearrow$ and $(y_n) \searrow$, (Example 3-2),

ii) $x_n < y_n \quad \forall n \in \mathbb{N}$, (Example 3-3),and

iii) $\lim(y_n - x_n) = \lim\left\{\left(1 + \frac{1}{n}\right)^{n+1} - \left(1 + \frac{1}{n}\right)^n\right\} =$

$\lim\left\{\left(1 + \frac{1}{n}\right)^n \left(1 + \frac{1}{n} - 1\right)\right\} = \lim\left\{\left(1 + \frac{1}{n}\right)^n \frac{1}{n}\right\} = \lim\left(1 + \frac{1}{n}\right)^n \lim\frac{1}{n} =$

$\ell \cdot 0 = 0.$

By Theorem 10-6, both sequences converge to the same limit, i.e. $\lim x_n = \lim y_n = \ell$.

Note: The common limit of (x_n) and (y_n) is called the number e, (**The Euler's number**),i.e.

$$\lim\left(1 + \frac{1}{n}\right)^n = \lim\left(1 + \frac{1}{n}\right)^{n+1} = e.$$

The number $e \cong 2.71828\cdots$, is **a transcendental number**, (proved by Charles Hermite in 1873).**A number is called transcendental, if it is not a root of any polynomial with integer coefficients**.

Another famous transcendental number is the number $\pi \cong 3.14159\cdots$ (proved by Von Lindemann, in 1882).

Example 10-2.

Find the limit of the sequence $y_n = \dfrac{1+\frac{1}{2}+\frac{1}{3}+\cdots+\frac{1}{n}}{n}$.

Solution

Let $x_n = \dfrac{1}{n}$. Obviously, the $\lim x_n = 0$. By Theorem 10-11,

$$\lim \frac{x_1+x_2+\cdots+x_n}{n} = 0 \Longrightarrow \lim y_n = \lim \frac{1+\frac{1}{2}+\cdots+\frac{1}{n}}{n} = 0.$$

Example 10-3.

Find the $\lim x_n$, if

$$x_n = \frac{1}{\sqrt{n^2+1}} + \frac{1}{\sqrt{n^2+2}} + \frac{1}{\sqrt{n^2+3}} + \cdots + \frac{1}{\sqrt{n^2+n}} .$$

Solution

We note that,

$$\sqrt{n^2+1} = \sqrt{n^2+1} < \sqrt{n^2+n} \Longrightarrow \frac{1}{\sqrt{n^2+1}} = \frac{1}{\sqrt{n^2+1}} > \frac{1}{\sqrt{n^2+n}}$$

$$\sqrt{n^2+1} < \sqrt{n^2+2} < \sqrt{n^2+n} \Longrightarrow \frac{1}{\sqrt{n^2+1}} > \frac{1}{\sqrt{n^2+2}} > \frac{1}{\sqrt{n^2+n}}$$

$$\vdots \qquad \vdots \qquad \vdots \qquad \vdots \qquad \vdots$$

$$\sqrt{n^2+1} < \sqrt{n^2+n} = \sqrt{n^2+n} \Longrightarrow \frac{1}{\sqrt{n^2+1}} > \frac{1}{\sqrt{n^2+n}} = \frac{1}{\sqrt{n^2+n}}$$

and adding term wise, we obtain,

$$\frac{n}{\sqrt{n^2+1}} > x_n > \frac{n}{\sqrt{n^2+n}} . \qquad\qquad (*)$$

Thus x_n is trapped between the two sequences, $b_n = \dfrac{n}{\sqrt{n^2+1}}$ and $a_n = \dfrac{n}{\sqrt{n^2+n}}$, both of which tend to the same limit 1, as $n \to \infty$, and from Theorem 10-4, we conclude that $\lim x_n = 1$.

Example 10-4.

If $x_n = \dfrac{8^n}{n!}$, show that $\lim x_n = 0$.

Solution

Let's apply Theorem 10-9. The ratio $\dfrac{x_{n+1}}{x_n} = \dfrac{8^{n+1}\, n!}{8^n\, (n+1)!} = \dfrac{8}{n+1}$ (see Problem 3-3), and since $\lim \dfrac{x_{n+1}}{x_n} = 0 < 1$, by virtue of Theorem 10-9, (the quotient rule), the $\lim x_n = 0$.

Example 10-5.

Find the $\lim \dfrac{n}{\sqrt[n]{n!}}$.

Solution

Let $x_n = \dfrac{n}{\sqrt[n]{n!}} = \sqrt[n]{\dfrac{n^n}{n!}} = \sqrt[n]{b_n}$, where $b_n = \dfrac{n^n}{n!}$.

By Theorem 10-10,

$$\lim x_n = \lim \sqrt[n]{b_n} = \lim \dfrac{b_{n+1}}{b_n}. \qquad (*)$$

However, $\dfrac{b_{n+1}}{b_n} = \dfrac{(n+1)^{n+1}/(n+1)!}{n^n/n!} = \dfrac{(n+1)^n}{n^n} = \left(1 + \dfrac{1}{n}\right)^n$, and since

$\lim \left(1 + \dfrac{1}{n}\right)^n = e$, (Example 10-1), from $(*)$ we have that the

$\lim x_n = \lim \sqrt[n]{b_n} = e$.

Example 10-6.

Consider the sequence $\{x_{n+1} = \sqrt{x_n + c}, \quad x_1 = \sqrt{c}, \ c > 0\}$.

a) Show that (x_n) is increasing and bounded above, and

b) Find the $\lim x_n$.

Solution

a) The given sequence (x_n) is increasing (see Problem 3-6) and bounded above by the number $c + 1$, (see Example 2-5). By Theorem 10-5,(**convergence of monotone and bounded sequences**),the sequence (x_n) converges to a limit $\ell \le c + 1$.

b) Having established that the limit of (x_n) exists and is finite, we have that the $\lim x_n = \lim x_{n+1} = \ell$, (by virtue of Theorem 10-7). Taking the limits of both sides of $x_{n+1} = \sqrt{x_n + c}$, we have,

$$\lim x_{n+1} = \lim \sqrt{x_n + c} \Rightarrow \lim x_{n+1} = \sqrt{\lim x_n + c} \Rightarrow \ell = \sqrt{\ell + c} \Rightarrow \ell^2 = \ell + c \Leftrightarrow \ell^2 - \ell - c = 0,$$

from which,

$$\ell_1 = \frac{1+\sqrt{1+4c}}{2} > 0, \quad \ell_2 = \frac{1-\sqrt{1+4c}}{2} < 0.$$

However, since all the terms of (x_n) are positive, **its limit should be non negative**, therefore ℓ_2 is rejected, and finally, $\lim x_n = \ell_1 = \frac{1+\sqrt{1+4c}}{2}$.

Example 10-7.

Show that the sequence $x_n = \frac{(-1)^n n}{3n+4}$ does not converge to a finite limit.

Solution

Let us consider the following two subsequences of (x_n),

$$y_n = x_{2n} = \frac{(-1)^{2n}(2n)}{3(2n) + 4} = \frac{2n}{6n + 4} \Rightarrow \lim y_n = \lim x_{2n} = \frac{2}{6}.$$

$$w_n = x_{2n+1} = \frac{(-1)^{2n+1}(2n + 1)}{3(2n + 1) + 4} = -\frac{2n + 1}{6n + 7} \Rightarrow \lim w_n = \lim x_{2n+1} = -\frac{2}{6}.$$

By virtue of Corollary 10-2, (**two subsequences of (x_n) converging to two different limits**), the sequence (x_n) has no limit.

Example 10-8.

Consider the sequence $\{\, x_{n+1} = \frac{1}{2}\left(x_n + \frac{1}{x_n}\right), \quad x_1 = c > 1\}$.

a) Show that (x_n) is decreasing and bounded, and

b) Find the $\lim x_n$.

Solution

a) The sequence (x_n) is decreasing and bounded,(see Problem 3-9),therefore (x_n) is convergent,(Theorem 10-5),and let $\lim x_n = \ell$, (a finite number).

b) The $\lim x_{n+1} = \lim x_n = \ell$, (by Theorem 10-7),and therefore,

$$\lim x_{n+1} = \lim \left\{\frac{1}{2}\left(x_n + \frac{1}{x_n}\right)\right\} \Rightarrow \lim x_{n+1} = \frac{1}{2}\left\{\lim x_n + \frac{1}{\lim x_n}\right\} \Rightarrow$$

$$\ell = \frac{1}{2}\left(\ell + \frac{1}{\ell}\right) \Leftrightarrow$$

$$\ell^2 = 1 \Rightarrow \{\ell = 1, \ or \ \ell = -1\}.$$

Since all the terms of (x_n) are positive, the negative root $\ell_2 = -1$ is rejected, and finally, the $\lim x_n = 1$.

Example 10-9.

Show that $\lim \sqrt[n]{n} = 1$.

Solution

By Theorem 10-10, we have,

$$\lim \sqrt[n]{n} = \lim \frac{n+1}{n} = 1.$$

Note: Similarly, it is easy to show that if p is any fixed positive number, $(p > 0)$ then $\lim \sqrt[n]{p} = 1$. The two limits, **$\lim \sqrt[n]{n} = 1$** and **$\lim \sqrt[n]{p} = 1, (p > 0)$**, are important limits, and appear quite often in more complicated sequences.

Example 10-10.

Let us consider the sequence,

$$x_n = \cos\left(\frac{x}{2}\right)\cos\left(\frac{x}{2^2}\right)\cos\left(\frac{x}{2^3}\right)\cdots\cos\left(\frac{x}{2^n}\right) = \prod_{k=1}^{n}\cos\left(\frac{x}{2^k}\right).$$

Show that,

a) $\lim x_n = \frac{\sin x}{x}$, and **b)** $\prod_{k=1}^{\infty}\left\{\cos\left(\frac{\pi}{2^{k+1}}\right)\right\} = \frac{2}{\pi}.$

Solution

a) In order to show part (a), the reader should know,(from elementary Calculus and Trigonometry),that

$$\lim_{y\to 0}\frac{\sin y}{y} = 1, \quad and \quad \cos x = \frac{\sin 2x}{2\sin x}.$$

We have,

$$x_n = \cos\left(\frac{x}{2}\right)\cos\left(\frac{x}{2^2}\right)\cos\left(\frac{x}{2^3}\right)\cdots\cos\left(\frac{x}{2^{n-1}}\right)\cos\left(\frac{x}{2^n}\right) \Rightarrow$$

$$x_n = \frac{\sin x}{2\sin\left(\frac{x}{2}\right)}\frac{\sin\left(\frac{x}{2}\right)}{2\sin\left(\frac{x}{2^2}\right)}\frac{\sin\left(\frac{x}{2^2}\right)}{2\sin\left(\frac{x}{2^3}\right)}\cdots\frac{\sin\left(\frac{x}{2^{n-2}}\right)}{2\sin\left(\frac{x}{2^{n-1}}\right)}\frac{\sin\left(\frac{x}{2^{n-1}}\right)}{2\sin\left(\frac{x}{2^n}\right)} \Rightarrow$$

$$x_n = \frac{1}{2^n}\frac{\sin x}{\sin\left(\frac{x}{2^n}\right)} = \frac{\left(\frac{x}{2^n}\right)}{\sin\left(\frac{x}{2^n}\right)}\frac{\sin x}{x}.$$

For any **fixed** x, the $\lim_{n\to\infty}\left(\frac{x}{2^n}\right) = 0$, while the term $\left(\frac{\sin x}{x}\right)$ does not depend on n, therefore,

$$\lim x_n = \frac{\sin x}{x}\cdot\lim\frac{\left(\frac{x}{2^n}\right)}{\sin\left(\frac{x}{2^n}\right)} = \frac{\sin x}{x}\cdot 1 = \frac{\sin x}{x}$$

and this completes the proof.

Note that $\lim\dfrac{\left(\frac{x}{2^n}\right)}{\sin\left(\frac{x}{2^n}\right)} = 1$, as $n\to\infty$, since $\left(\frac{x}{2^n}\right)\to 0$, as $n\to\infty$.

b) From part (a),

$$\lim x_n = \frac{\sin x}{x} \Rightarrow \prod_{k=1}^{\infty}\cos\left(\frac{x}{2^k}\right) = \frac{\sin x}{x}.$$

This equation is **an identity, true for all** x, therefore it will be true for $x = \frac{\pi}{2}$ as well. In this case,

$$\prod_{k=1}^{\infty} \cos\left(\frac{\pi}{2^{k+1}}\right) = \frac{\sin\left(\frac{\pi}{2}\right)}{\left(\frac{\pi}{2}\right)} = \frac{1}{\left(\frac{\pi}{2}\right)} = \frac{2}{\pi}.$$

PROBLEMS

10-1) If $x_n = \dfrac{n}{\sqrt[n]{(n+1)(n+2)(n+3)\cdots(2n)}}$, show that $\lim x_n = \dfrac{e}{4}$.

Hint: Apply Theorem 10-10.

10-2) If $x_n = \sqrt[n]{3^n + 4^n + 5^n}$, show that $\lim x_n = 5$.

Hint: $5^n < 3^n + 4^n + 5^n < 5^n + 5^n + 5^n$, i.e. $5 < x_n < 5\sqrt[n]{3}$. Recall that $\lim \sqrt[n]{3} = 1$, (see Example 10-9) and then apply Theorem 10-4.

10-3) Let $\{x_{n+1} = \sqrt{cx_n}, \ x_1 = 1\}$, where c is a constant > 1.

a) Show that (x_n) is increasing and bounded above, by the number c.

b) Show that $\lim x_n = c$.

10-4) Show that the sequence $y_n = \dfrac{n^2}{7^n}$ is bounded.

Hint: It suffices to show that (y_n) converges and then apply Theorem 10-2. To show that (y_n) converges, apply Theorem 10-9.

10-5) Find the limit of the following sequences,

a) $x_n = \dfrac{5^n \cdot n!}{(8n)^n}$,   **b)** $y_n = \left(\sqrt{n+5} - \sqrt{n}\right) \cdot \sqrt{n+9}$.

(Answer: **a)** 0, **b)** $\dfrac{5}{2}$).

10-6) Let $\left\{x_{n+1} = \frac{5+\sqrt{4x_n-7}}{2}, \quad x_1 = 2\right\}$. Show that (x_n) is increasing and bounded above, and that $\lim x_n = 4$.

10-7) Making use of Theorem 10-12, find the limit of the following sequences,

a) $y_n = \frac{1+\sqrt{2}+\sqrt[3]{3}+\sqrt[4]{4}+\cdots+\sqrt[n]{n}}{n}$, **b)** $z_n = (\sqrt[n]{1})(\sqrt[2n]{2})(\sqrt[3n]{3})(\sqrt[4n]{4})\cdots(\sqrt[n^2]{n})$, and

c) $w_n = \dfrac{n}{1+\frac{1}{\sqrt{2}}+\frac{1}{\sqrt[3]{3}}+\frac{1}{\sqrt[4]{4}}+\cdots+\frac{1}{\sqrt[n]{n}}}$.

Hint: If $x_n = \sqrt[n]{n}$, the $\lim x_n = 1$,(see Example 10-9).

(Answer: All sequences converge to the number 1).

10-8) Let (x_n) and (y_n) be two sequences defined as,

$\left\{x_{n+1} = \sqrt{x_n y_n}, \quad x_1 = a \quad and \quad y_{n+1} = \frac{x_n+y_n}{2}, \quad y_1 = b, \quad 0 < a < b\right\}$.

Show that,

a) $x_n < y_n$, $\forall n \in \mathbb{N}$,

b) (x_n) is increasing and (y_n) is decreasing,

c) $\lim(y_n - x_n) = 0$.

Hint: To show part (c), show that $0 < y_{k+1} - x_{k+1} < \frac{y_k-x_k}{2}$, apply for $k = 1,2,3,\cdots(n-1)$ and multiply term wise, to get, $0 < y_n - x_n < \frac{b-a}{2^{n-1}}$, etc.

10-9) Show that the following sequences, have no limit,

a) $x_n = \frac{1+(-1)^n}{5}$, **b)** $y_n = \sin\frac{n\pi}{2}$, **c)** $z_n = \frac{(-1)^n n^3}{4n^3+7}$.

10-10) Show that $\lim \dfrac{\sqrt[n]{1^n+2^n+3^n+\cdots+n^n}}{n} = 1$.

Hint: $n^n < 1^n + 2^n + 3^n + \cdots + n^n < n\cdot n^n$.

10-11) If $y_n = \prod_{k=2}^{n}\left(1-\frac{1}{k^2}\right)$, show that $\lim y_n = \frac{1}{2}$.

10-12) If $w_n = \prod_{k=1}^{n}\left(1 + x^{2^k}\right)$, where $|x| < 1$, show that $\lim w_n = \frac{1}{1-x^2}$.

Hint: $1 + x^{2^k} = \left(1 - x^{2^{(k+1)}}\right) \div \left(1 - x^{2^k}\right)$, $k \in \mathbb{N}$.

10-13) Consider the sequences (x_n) and (y_n), defined as,

$$\left\{x_{n+1} = \frac{x_n + y_n}{2}, \quad x_1 = k \quad and \quad y_{n+1} = \frac{x_{n+1} + y_n}{2}, \quad y_1 = \ell, \quad 0 < \ell < k\right\}.$$

Show that

a) (x_n) is increasing and (y_n) is decreasing,

b) $y_n < x_n \quad \forall n \in \mathbb{N}$,

c) $\lim(y_n - x_n) = 0$.

10-14) Show that the sequence $w_n = \frac{n^3 \cos n + n^2 \sin n}{n^4 + 2}$ is bounded.

Hint: It suffices to show that (w_n) converges, and then apply Theorem 10-2.

10-15) If (x_n) is an increasing sequence of positive terms, converging to a finite limit ℓ, show that $\lim \sqrt[n]{x_n} = 1$.

Hint: Assume that $\lim x_n = \ell$, and then apply Theorem 10-10.

10-16) Consider the sequence $x_n = \frac{1}{1^2} + \frac{1}{2^2} + \frac{1}{3^2} + \cdots + \frac{1}{n^2}$ and show that (x_n) is convergent.

Hint: The sequence (x_n) is increasing and bounded, $(0 < x_n < 2)$. To prove this consider the inequality, $\frac{1}{n^2} < \frac{1}{n-1} - \frac{1}{n}$, $n = 2,3,4,\cdots$. By Theorem 10-5 the sequence (x_n) converges to a limit $\ell \leq 2$.

Note: It can be shown that,

$$\lim x_n = \frac{1}{1^2} + \frac{1}{2^2} + \frac{1}{3^2} + \cdots = \sum_{n=1}^{\infty}\frac{1}{n^2} = \frac{\pi^2}{6},$$

an unexpected result. This was the famous **Basel Problem** proved for the first time by **Leonard Euler**, one of the greatest Mathematicians of all times.

10-17) If $\{ x_{n+1} = \frac{1}{2}\left(x_n + \frac{c}{x_n}\right), \quad c > 0, \; x_1 > 0, \; x_1^2 > c\}$, show that

a) (x_n) is decreasing and bounded below, and

b) $\lim x_n = \sqrt{c}$.

10-18) Consider the sequence $\{ x_{n+1} = 1 + \frac{x_n}{1+x_n}, \quad x_1 = c > 0\}$, (see Example 3-8),and show that $\lim x_n = \frac{1+\sqrt{5}}{2}$.

10-19) For the sequence (x_n) defined in Problem 2-4,show that it is increasing and find its limit.

(Answer: $\lim x_n = d$).

10-20) Consider the sequence $\{ x_{n+1} = \sqrt{1 - x_n + x_n^2}, \quad 0 < x_1 < 1\}$. Show that (x_n) is increasing, bounded, and that $\lim x_n = 1$.

10-21) If $x_n = (-1)^n \frac{n^3}{3^n}$, show that $\lim x_n = 0$.

Hint: Apply the quotient rule.

10-22) Let $x_1, x_2, x_3, \cdots, x_n$ be n positive numbers. The arithmetic mean (A.M), the geometric mean (G.M) and the harmonic mean (H.M) of these numbers, are defined as,

$$A.M = \frac{x_1 + x_2 + \cdots x_n}{n}, \quad G.M = \sqrt[n]{x_1 x_1 \cdots x_1}, \quad H.M = \frac{n}{\frac{1}{x_1} + \frac{1}{x_2} + \cdots + \frac{1}{x_n}}.$$

These three means satisfy the **Cauchy's inequality,** $H.M \le G.M \le A.M$. (*)

Making use of Theorem 10-11, show that the sequences (y_n) and (w_n), as defined in Theorem 10-12, tend to ℓ, and then using (*) and Theorem 10-4,show that the sequence (z_n), in Theorem 10-12, also tends to ℓ.

10-23) Prove Theorem 10-11, in case $\ell = +\infty$.

10-24) If $(y_n) = \sqrt[n]{n!}$, show that $\lim y_n = +\infty$.

Hint: Apply Theorem 10-10.

10-25) If $r > 0$, and $x_n = \sqrt[n]{\dfrac{(r+1)(r+2)(r+3)\cdots(r+n)}{n!}}$, show that $\lim x_n = 1$.

10-26) If $x_n = \sqrt[n]{\dfrac{3n^2+7}{n^3+5n+1}}$, show that $\lim x_n = 1$.

10-27) If $p > 0$, and $x_n = \sqrt[n]{p^n + p^{-n}}$, show that $\lim x_n = \left\{ \begin{array}{ll} p & if \ \ p \geq 1 \\ \frac{1}{p} & if \ \ 0 < p < 1 \end{array} \right\}$.

10-28) If $y_n = \dfrac{3\sqrt[n]{n}+7\sqrt[3]{n}}{2n}$, show that $\lim y_n = 0$.

Hint: Since $\lim \sqrt[n]{n} = 1$, the sequence $\left(\sqrt[n]{n}\right)$ is bounded,(see Example 9-2), while the sequence $\left(\dfrac{1}{2n}\right)$ is null, therefore the sequence $\left(\dfrac{\sqrt[n]{n}}{2n}\right)$ is null,(see Problem 9-11),etc.

10-29) If $x_n = \dfrac{2^3-1}{2^3+1} \cdot \dfrac{3^3-1}{3^3+1} \cdot \dfrac{4^3-1}{4^3+1} \cdots \dfrac{n^3-1}{n^3+1}$, find the limit of (x_n).

(Answer: $\dfrac{2}{3}$**).**

10-30) If $y_n = \dfrac{x^n}{n!}$, show that $\lim y_n = 0 \ \ \forall x \in \mathbb{R}$.

10-31) Let us consider the sequence $\{3x_{n+1} = 1 + x_n^2 , \ x_1 > 0\}$.For what values of x_1 the sequence (x_n) is decreasing? What is the $\lim(x_n)$?

(Answer: $\dfrac{3-\sqrt{5}}{2} < x_1 < \dfrac{3+\sqrt{5}}{2}, \ \ \lim x_n = \dfrac{3-\sqrt{5}}{2}$**).**

10-32) Show that the sequence $y_n = \dfrac{1}{1^4+1} + \dfrac{2}{2^4+1} + \dfrac{3}{3^4+1} + \cdots + \dfrac{n}{n^4+1}$ is convergent.

10-33) If $x \geq 1$, show that $\lim \dfrac{x^{-(n+1)}+x^{(n+1)}}{x^{-n}+x^n} = x$. What is the limit if $0 < x < 1$?

10-34) Show that $\lim \dfrac{7n+3\sqrt[3]{n}+5}{n^2+\sqrt[n]{n}} = 0.$

10-35) Show that $\lim \dfrac{1.7^n n!}{n^n} = 0.$

10-36) Making use of the quotient rule, show that $x_n = n^p q^n$ where $p > 0$ and $|q| < 1$ is a null sequence.

10-37) If $\lim(x_n^2 + y_n^2) = 0$, show that $\lim x_n = 0$ and $\lim y_n = 0$. (Assume that (x_n) and (y_n) are sequences of real numbers).

10-38) Show that the sequence $y_n = \dfrac{\sin \sqrt[n]{n} + \cos \sqrt[5]{n}}{\sqrt[3]{n^4}}$ is null.

10-39) Consider the sequence $\left\{ x_{n+1} = \dfrac{9+x_n^2}{2x_n} \,,\ x_1 = 1 \right\}$, and show that (x_n) is decreasing and bounded below by the number 3, and that $\lim x_n = 3.$

10-40) If (x_n) is a decreasing sequence of positive terms convergent to a limit $\ell > 0$, show that $\lim \sqrt[n]{x_n} = 1$. Show the same if (x_n) is an increasing sequence of positive terms convergent to a limit $\ell > 0$.

Hint: If (x_n) is decreasing, then $x_1 > x_2 > \cdots > x_n > \cdots \geq l > 0$, and $\sqrt[n]{x_1} > \sqrt[n]{x_n} > \sqrt[n]{\ell}$, and since $\sqrt[n]{x_1} \to 1$, $\sqrt[n]{\ell} \to 1$, as $n \to \infty$, by Theorem 10-4, the in between trapped sequence $\sqrt[n]{x_n}$ will be convergent to the same limit.

11. The Number e and Other Remarkable Limits.

In Example 10-1, **the Euler's number e**, was defined as

$$e = \lim_{n \to \infty} \left(1 + \frac{1}{n}\right)^n = \lim_{n \to \infty} \left(1 + \frac{1}{n}\right)^{n+1}. \qquad (11\text{-}1)$$

Let us now consider the sequence $w_n = \left(1 - \frac{1}{n}\right)^n, \quad n = 2, 3, 4, \cdots$

Obviously,

$$w_{n+1} = \left(1 - \frac{1}{n+1}\right)^{n+1} = \left(\frac{n}{n+1}\right)^{n+1} = \left(\frac{1}{1+\frac{1}{n}}\right)^{n+1} = \frac{1}{\left(1+\frac{1}{n}\right)^n} \cdot \frac{1}{1+\frac{1}{n}} \cdot \quad (*)$$

As $n \to \infty$,

$$\frac{1}{\left(1+\frac{1}{n}\right)^n} \to e \quad and \quad \frac{1}{1+\frac{1}{n}} = 1,$$

therefore from (*),

$\lim_{n \to \infty} w_{n+1} = \frac{1}{e} \cdot 1 = \frac{1}{e}$, and since $\lim w_{n+1} = \lim w_n$, (Theorem 10-7),

$$\lim w_n = \lim_{n \to \infty} \left(1 - \frac{1}{n}\right)^n = \frac{1}{e}. \qquad (11\text{-}2)$$

Equations (11-1) and (11-2), are special cases of the following more general Theorem:

Theorem 11-1.
If (x_n) is a sequence of positive terms, diverging to $+\infty$, $(\lim x_n = +\infty)$, then

$$\lim_{n \to \infty} \left(1 + \frac{1}{x_n}\right)^{x_n} = e, \quad and \quad \lim_{n \to \infty} \left(1 - \frac{1}{x_n}\right)^{x_n} = \frac{1}{e}. \qquad (11\text{-}3)$$

Equations (11-1) and (11-2), are obtained from (11-3), if $x_n = n$.

Proof: For the proof, see Example 11-8.

If $\lim_{n\to\infty} x_n = +\infty$, then the sequence $y_n = \frac{1}{x_n}$, will be a null sequence, ($\lim_{n\to\infty} y_n = 0$), and Theorem 11-1, may be stated equivalently, as

Theorem 11-2.

If (y_n) is a null sequence, $(\lim_{n\to\infty} y_n = 0,\ y_n \neq 0\ and\ y_n = -1, \forall n \in \mathbb{N})$, then

$$\lim_{n\to\infty}(1+y_n)^{\frac{1}{y_n}} = e,\ and\ \lim_{n\to\infty}(1-y_n)^{\frac{1}{y_n}} = \frac{1}{e}. \qquad (11\text{-}4)$$

Let us now examine the sequence $b_n = \left(1+\frac{c}{n}\right)^n$, $c > 0$. Then,

$$b_n = \left\{\left(1+\frac{1}{\left(\frac{n}{c}\right)}\right)^{\frac{n}{c}}\right\}^c = \left\{\left(1+\frac{1}{x_n}\right)^{x_n}\right\}^c,\ where\ x_n = \frac{n}{c},\ (\lim_{n\to\infty} x_n = +\infty).$$

By Theorem (11-1),

$$\lim_{n\to\infty} b_n = \lim_{n\to\infty}\left(1+\frac{c}{n}\right)^n = e^c.$$

Similarly, one may show that,

$$\lim_{n\to\infty}\left(1-\frac{c}{n}\right)^n = e^{-c}.$$

We have thus shown that,

$$\lim_{n\to\infty}\left(1+\frac{c}{n}\right)^n = e^c\ and\ \lim_{n\to\infty}\left(1-\frac{c}{n}\right)^n = e^{-c}, c > 0. \qquad (11\text{-}5)$$

In particular, if x is any real number, the exponential function e^x can be defined as,

$$e^x = \lim\left(1+\frac{x}{n}\right)^n,\ -\infty < x < +\infty.$$

It can be shown that the Euler's number e, has another **series representation**, which we state without proof.

$$e = \lim \left(1 + \frac{1}{n}\right)^n = 1 + \frac{1}{1!} + \frac{1}{2!} + \frac{1}{3!} + \frac{1}{4!} + \frac{1}{5!} + \cdots = \sum_{n=0}^{\infty} \frac{1}{n!} \cdot \quad (0! = 1). \quad (11\text{-}6)$$

$$e^{-1} = \lim \left(1 - \frac{1}{n}\right)^n = 1 - \frac{1}{1!} + \frac{1}{2!} - \frac{1}{3!} + \frac{1}{4!} - \frac{1}{5!} + \cdots = \sum_{n=0}^{\infty} \frac{(-1)^n}{n!} \cdot \quad (11\text{-}7)$$

For the rest of this chapter, the reader is supposed to be familiar with **the exponential function $y = e^x$, $-\infty < x < \infty$, the logarithmic function $y = \ln x$, $0 < x < \infty$, and with elements from the differential calculus, in particular with the so called De L'Hospital Rule.**

The following important Theorem connects continuous functions with limits of sequences.

Theorem 11-3.

If a function $y = f(x)$ is continuous at $x = \ell$, and (x_n) is a sequence with $\lim x_n = \ell$, then $\lim\{f(x_n)\} = f(l) = f(\lim x_n) \cdot$ (11-8)

Proof: Since $f(x)$ is assumed to be continuous at $x = \ell$,

$$\forall \varepsilon > 0 \quad \exists \delta = \delta(\varepsilon): |f(x) - f(\ell)| < \varepsilon \ \text{whenever} \ |x - \ell| < \delta \cdot \quad (*)$$

Now let us choose N such that $|x_n - \ell| < \delta, \quad \forall n > N$.

Then from (*) we have,

$$|f(x_n) - f(\ell)| < \varepsilon, \ \text{whenever} \ n > N, \text{meaning that} \quad \lim\{f(x_n)\} = f(l) = f(\lim x_n)$$ and this completes the proof.

We remind the reader that **the functions $y = e^x$ and $y = \ln x$, are continuous at every point, within their interval of definition.**

We may now evaluate some special limits, involving the logarithmic function.

Theorem 11-4.
The sequences,

$$x_n = \frac{\ln n}{n}, y_n = \frac{\ln n}{n^p}, and \ z_n = \frac{(\ln n)^q}{n^p}$$

where p and q are arbitrary positive numbers, are null sequences.

Proof: The function $y = \dfrac{\ln x}{x}$ approaches zero as $x \to \infty$, i.e. $\lim_{x \to \infty} \dfrac{\ln x}{x} = 0$,

(proof easy using the De Lhospital Rule). If we imagine that x approaches infinity,

through the sequence $(a_n) = (n)$, we obtain $\lim x_n = \lim \dfrac{\ln n}{n} = 0.$

Similarly, the function $y = \dfrac{\ln x}{x^p}$, $(p > 0)$, tends to zero, as $x \to \infty$, and

reasoning as before, the $\lim y_n = \lim \dfrac{\ln n}{n^p} = 0.$

Finally, $z_n = \dfrac{(\ln n)^q}{n^p} = \left\{ \dfrac{\ln n}{n^{\left(\frac{p}{q}\right)}} \right\}^q$, and since $\dfrac{\ln n}{n^{\left(\frac{p}{q}\right)}}$ is null, (z_n) will be null as well.

Another special limit is the so called, **Euler-Mascheroni constant** γ, defined as

$$\gamma = \lim_{n \to \infty} \left(1 + \frac{1}{2} + \frac{1}{3} + \frac{1}{4} + \frac{1}{5} + \cdots + \frac{1}{n} - \ln n \right) \cong 0.57721 \ldots \ (11\text{-}9)$$

A thorough treatment on the constant γ and on the series representation of the Euler's number e will be given in our next e book, on Infinite Series and Products, to appear soon.

We conclude this chapter, with two important limits, already mentioned in Example 10-9,

$$\lim_{n \to \infty} \sqrt[n]{p} = 1 , (p > 0), \quad and \quad \lim_{n \to \infty} \sqrt[n]{n} = 1. \quad (11\text{-}10)$$

For an alternative proof, see Problem 11-5.

Example 11-1.

If $x_n = \left(1 + \dfrac{2}{3n} \right)^n$, find the $\lim x_n$.

Solution

The general term of the given sequence is $x_n = \left(1 + \dfrac{2}{3n} \right)^n = \left(1 + \dfrac{\left(\frac{2}{3}\right)}{n} \right)^n$,

and $\lim x_n = e^{\left(\frac{2}{3}\right)} = \sqrt[3]{e^2}$, (from (11-5), with $c = \dfrac{2}{3}$).

Example 11-2.

If $y_n = \left(1 - \dfrac{1}{3n}\right)^n$, find the $\lim y_n$.

Solution

The sequence $y_n = \left(1 - \dfrac{1}{3n}\right)^n = \left(1 - \dfrac{\left(\frac{1}{3}\right)}{n}\right)^n$, and making use of equation (11-5),

we have $\lim y_n = e^{\left(-\frac{1}{3}\right)} = \dfrac{1}{\sqrt[3]{e}}$.

Example 11-3.

If $w_n = \left(1 - \dfrac{1}{n^2}\right)^n$, $n = 2,3,4,\cdots$, find the $\lim w_n$.

Solution

$$w_n = \left(1 - \dfrac{1}{n^2}\right)^n = \left\{\left(1 + \dfrac{1}{n}\right) \cdot \left(1 - \dfrac{1}{n}\right)\right\}^n = \left(1 + \dfrac{1}{n}\right)^n \cdot \left(1 - \dfrac{1}{n}\right)^n,$$

and as $n \to \infty$, $\lim w_n = e \cdot e^{-1} = 1$.

Example 11-4.

Find the limit of the sequence $x_n = \dfrac{(3n+1) \cdot (2n+4) \cdot n^n}{(n+1)^{n+2}}$.

Solution

$$x_n = \left(\dfrac{n}{n+1}\right)^n \cdot \dfrac{(3n+1) \cdot (2n+4)}{(n+1)^2} = \left(\dfrac{1}{1+\frac{1}{n}}\right)^n \cdot \dfrac{(3n+1) \cdot (2n+4)}{(n+1)^2},$$

and as $n \to \infty$, $\lim x_n = \dfrac{1}{e} \cdot 6 = \dfrac{6}{e}$. (See Example 9-7).

Example 11-5.

Find the limit of the sequence $x_n = \left(1 + \dfrac{3}{\sqrt{n}}\right)^{\sqrt{n}}$.

Solution

The general term of the given sequence can be expressed as $x_n = \left\{ \left(1 + \right. \right.$

$$\left. \left. \frac{1}{\left(\frac{\sqrt{n}}{3}\right)} \right)^{\left(\frac{\sqrt{n}}{3}\right)} \right\}^3,$$

and since $\frac{\sqrt{n}}{3} \to \infty$, as $n \to \infty$, the $\lim x_n = e^3$, (See Theorem 11-1 with $x_n = \frac{\sqrt{n}}{3}$).

Example 11-6.

If $x_n = \left(1 + \frac{1}{2n}\right)^n$, $y_n = \frac{\ln n}{\sqrt{n}}$, find the $\lim(x_n \cdot y_n)$.

Solution

From (11-5), $\lim x_n = e^{\frac{1}{2}} = \sqrt{e}$, while from Theorem 11-4, $\lim y_n = 0$, therefore,

$$\lim(x_n \cdot y_n) = (\lim x_n) \cdot (\lim y_n) = \sqrt{e} \cdot 0 = 0.$$

Example 11-7.

If $x_n = \left(1 + \frac{\ln n}{\sqrt{n}}\right)^{\left(\frac{\sqrt{n}}{\ln n}\right)}$, find the $\lim x_n$.

Solution

The sequence $y_n = \frac{\ln n}{\sqrt{n}}$ is null,(Theorem 11-4, with $p = \frac{1}{2}$), so $x_n = (1 + y_n)^{\left(\frac{1}{y_n}\right)}$,where y_n is a null sequence. According to Theorem 11-2, $\lim x_n = e$.

Example 11-8.

Prove Theorem 11-1.

Solution

Let (x_n) be a sequence of positive terms, diverging to $+\infty$, $(\lim x_n = +\infty)$. If we call $[x_n]$ the integer part of x_n, then $[x_n] \leq x_n < [x_n] + 1$, (see equation (5-5)), or if we define the sequence $y_n = [x_n]$, $y_n \leq x_n < y_n + 1$.   (*)

The sequence y_n is an increasing sequence of positive integers, diverging to $+\infty$, i.e. $\lim y_n = +\infty$. In terms of y_n, inequality (*) becomes,

$$y_n \leq x_n < y_n + 1 \Rightarrow \frac{1}{y_n} \geq \frac{1}{x_n} > \frac{1}{y_n + 1} \Rightarrow 1 + \frac{1}{y_n} \geq 1 + \frac{1}{x_n} > 1 + \frac{1}{y_n + 1}$$

$$\Rightarrow \left(1 + \frac{1}{y_n}\right)^{x_n} \geq \left(1 + \frac{1}{x_n}\right)^{x_n} > \left(1 + \frac{1}{y_n + 1}\right)^{x_n},$$

or taking (*) into consideration,

$$\left(1 + \frac{1}{y_n}\right)^{y_n+1} \geq \left(1 + \frac{1}{x_n}\right)^{x_n} > \left(1 + \frac{1}{y_n+1}\right)^{y_n}.$$ (**)

Since (y_n) is a subsequence of the sequence (n) of the positive integers, **the sequence $\left(1 + \frac{1}{y_n}\right)^{y_n+1}$ is a subsequence of $\left(1 + \frac{1}{n}\right)^{n+1}$**, which tends to the number e, and **the sequence $\left(1 + \frac{1}{y_n+1}\right)^{y_n}$ is a subsequence of $\left(1 + \frac{1}{n+1}\right)^{n}$** $= \left(1 + \frac{1}{n+1}\right)^{n+1} \cdot \left(1 + \frac{1}{n+1}\right)^{-1}$, which again tends to e, as $n \to \infty$. **The sequence $\left(1 + \frac{1}{x_n}\right)^{x_n}$ is trapped between two sequences,(see equation (*)), each one of which tends to the same limit e**, and therefore, by virtue of Theorem 10-4,(Trapped sequences), $\lim \left(1 + \frac{1}{x_n}\right)^{x_n} = e$, and this completes the proof.

To prove that $\lim \left(1 - \frac{1}{x_n}\right)^{x_n} = \frac{1}{e}$, we note that

$$\left(1 - \frac{1}{x_n}\right)^{x_n} = \left(\frac{x_n-1}{x_n}\right)^{x_n} = \frac{1}{\left(\frac{x_n}{x_n-1}\right)^{x_n}} = \frac{1}{\left(1+\frac{1}{x_n-1}\right)^{x_n-1}} \cdot \frac{1}{1+\frac{1}{x_n-1}},$$

and $\lim \left(1 - \frac{1}{x_n}\right)^{x_n} = \lim \frac{1}{\left(1+\frac{1}{x_n-1}\right)^{x_n-1}} \cdot \lim \frac{1}{1+\frac{1}{x_n-1}} = \frac{1}{e} \cdot 1 = \frac{1}{e}.$

Example 11-9.

If $0 < x < 1$, show that $\lim\left(x + \frac{1}{n}\right)^n = 0.$

Solution

$$\lim\left(x + \frac{1}{n}\right)^n = \lim\left\{x^n \cdot \left(1 + \frac{1}{x \cdot n}\right)^n\right\} = \lim\left\{x^n \cdot \left(1 + \frac{\left(\frac{1}{x}\right)}{n}\right)^n\right\}$$

$$= \lim x^n \cdot \lim\left(1 + \frac{\left(\frac{1}{x}\right)}{n}\right)^n = 0 \cdot e^{1/x} = 0.$$

Note that since $0 < x < 1$, the $\lim x^n = 0$, (see Theorem 8-1).

PROBLEMS

11-1) Find the limit of the following sequences,

$$a)\ x_n = \left(1 + \frac{1}{2n}\right)^n, \quad b)\ y_n = \left(1 - \frac{4}{n^2}\right)^n, \quad c)\ w_n = \left(1 - \frac{1}{n+1}\right)^{n+5}.$$

(Answer: $a)\sqrt{e},\ b)1,\ c)e^{-1}$).

11-2) Let (x_n) be a sequence defined as,

$$x_n = 1 + \frac{1}{1!} + \frac{1}{2!} + \frac{1}{3!} + \frac{1}{4!} + \cdots + \frac{1}{n!}.$$

Show that (x_n) is increasing, and bounded above by the number 3.(According to Theorem 10-5, the sequence (x_n) converges. **The $\lim x_n = e$** , (see (11-6)).

Hint: To show that $(x_n) \nearrow$ is easy. To show that (x_n) is bounded above, show first that $\frac{1}{n!} < \frac{1}{2^{n-1}}$, $n = 2,3,4,\cdots$

11-3) Find the limit of $x_n = \frac{(n+1)^n}{\frac{(n^2-1)}{n}}$.

(Answer: e).

11-4) Show that $\lim \left(\frac{\sqrt[n]{a}+\sqrt[n]{b}}{2}\right)^n = \sqrt{a \cdot b}$, $a > 0, b > 0$.

Hint: Consider the function

$$y = \left(\frac{a^{\left(\frac{1}{x}\right)} + b^{\left(\frac{1}{x}\right)}}{2}\right)^x,$$

and find the $\lim_{x \to \infty} y$. This will be the sought for limit of the given sequence.

$$\ln y = x \cdot \left\{\ln\left(a^{\left(\frac{1}{x}\right)} + b^{\left(\frac{1}{x}\right)}\right) - \ln 2\right\},$$

and $\lim_{x \to \infty} \ln y = \lim_{u \to 0} \frac{\ln(a^u + b^u) - \ln 2}{u} = \frac{0}{0}$, which may be evaluated by the De Lhospital Rule, etc.

11-5) Consider the functions $y_1(x) = x^{\frac{1}{x}}$, $x > 0$, and $y_2(x) = p^{\frac{1}{x}}$, $p > 0$, $x > 0$.

Show that $\lim_{x \to \infty} y_1(x) = 1$, $\lim_{x \to \infty} y_2(x) = 1$, and then infer that $\lim \sqrt[n]{n} = 1$, and $\lim \sqrt[n]{p} = 1$.

11-6) If $x_n = n\left(\sqrt[n]{p} - 1\right)$, find the $\lim x_n$, $(p > 0)$.

Hint: Consider the function $y = x \ln\left(p^{\frac{1}{x}} - 1\right)$ and find the $\lim_{x \to \infty} y$.

(Answer: $\ln p$).

11-7) If $x_n = 1 + \frac{1}{2} + \frac{1}{3} + \frac{1}{4} + \frac{1}{5} + \cdots + \frac{1}{n} - \ln n$, show that the sequence (x_n) is decreasing and bounded below,(and therefore converges to a limit, which is called γ, (see (11-9))).

11-8) Make use of (11-6) to approximate e, to three decimal place accuracy.

11-9) Make use of (11-7) to approximate e^{-1}, to three decimal place accuracy.

(Answer: 0.367...).

11-10) Make use of (11-9) to approximate γ to two decimal place accuracy.

11-11) Find the limit of the following sequences,

a) $x_n = \sin\left(1 + \frac{1}{n}\right)^n$, *b)* $y_n = e^{\sqrt[n]{n}}$, *c)* $w_n = n \cdot \ln\left(1 - \frac{1}{2n}\right)$, *d)* $z_n = n \cdot \ln\left(1 + \frac{3}{n}\right)$.

Hint: Apply Theorem 11-3.

(Answer: **a)** $\sin e$, **b)** e, **c)** $-\frac{1}{2}$, **d)** 3).

11-12) If $\lim x_n = +\infty$, $\lim y_n = +\infty$, and $\lim (x_n - y_n) = c$, (constant), show that $\lim\left(\frac{x_n}{y_n}\right) = 1$.

11-13) Making use of the previous problem, show that $\lim\frac{\ln(n+p)}{\ln n} = 1$, where p is any positive constant.

11-14) If $x_n = \left(\frac{1}{3} + \frac{5}{n}\right)^n$, show that $\lim x_n = 0$.

Hint: $x_n = \left(\frac{1}{3} + \frac{5}{n}\right)^n = \left\{\frac{1}{3}\left(1 + \frac{15}{n}\right)\right\}^n = \left(\frac{1}{3}\right)^n\left(1 + \frac{15}{n}\right)^n$, etc.

11-15) Find the limits of the following sequences

a) $x_n = \frac{\sqrt{(\ln n)^3}}{\sqrt[3]{n}}$, **b)** $y_n = \frac{1}{n}\ln\frac{(n+1)(n+2)(n+3)\cdots(2n)}{n^n}$.

(Answer: **a)** 0, **b)** $\ln 4 - 1$).

11-16) Show that $\lim \left(1 - \frac{1}{5n}\right)^{n^2+3n} = 0.$

11-17) Show that $\lim \left(1 + \frac{1}{3n^2}\right)^{4n} = 0.$

11-18) Show that $\lim \left(\frac{5n+4}{5n+2}\right)^{3n} = \sqrt[5]{e^6}.$

12. Recursive Sequences.

In Chapter 1 we have mentioned that a sequence (x_n) is well defined if:

a) The nth term $x_n = f(n)$ is **a known function of n**, from which

$$x_1 = f(1), \; x_2 = f(2), x_3 = f(3) \cdots, x_n = f(n), \cdots \cdots \quad \text{or}$$

b) The $(n+1)^{th}$ term of the sequence is expressed in terms of its predecessor, i.e.

$$x_{n+1} = f(x_n), \quad x_1 = c \text{ (given)}. \tag{12-1}$$

Equation (12-1) defines a **first order recursive sequence**. The terms of the sequence are completely specified, $x_1 = c, \; x_2 = f(x_1), \; x_3 = f(x_2), \cdots \cdots$

c) The $(n+2)^{th}$ term of the sequence is expressed in terms of its two predecessors, i.e.

$$x_{n+2} = f(x_{n+1}, x_n), \quad x_1 = c_1, \; x_2 = c_2, \quad \text{(given)}. \tag{12-2}$$

This is **a second order recursive sequence**, where,

$$x_1 = c_1, \; x_2 = c_2, \; x_3 = f(x_2, x_1), \; x_4 = f(x_3, x_2), \cdots \cdots$$

Higher order recursive sequences are similarly defined, (see Problem 12-13).

One of the main problems in recursive sequences, **is to express the nth term x_n in terms of** , i.e. to find the **functional dependence of x_n from n**. We shall demonstrate the approach towards this problem, by considering various characteristic Examples.
Prior to doing so, we shall remind the reader, the definition for **the arithmetic and the geometric progression.**

We say that the sequence (x_n) is **an arithmetic progression**, if

$$x_{n+1} = d + x_n, \; x_1 = c \text{ , (given)}. \tag{12-3}$$

This means that

$$x_1 = c, \quad x_2 = c + d, \quad x_3 = c + 2d, \quad x_4 = c + 3d, \quad \cdots, x_{n+1} = c + nd. \qquad (12\text{-}4)$$

The number $x_1 = c$ is the first term while the number d is the ratio of the arithmetic progression.

The sequence (y_n) is **a geometric progression**, if

$$x_{n+1} = d \cdot x_n, \quad x_1 = c, (c \neq 0), \textbf{(given)}. \qquad (12\text{-}5)$$

This means that

$$x_1 = c, \quad x_2 = c \cdot d, \quad x_3 = c \cdot d^2, \quad x_4 = c \cdot d^3, \quad \cdots x_{n+1} = c \cdot d^n. \quad (12\text{-}6)$$

The number $x_1 = c$ is the first term, while the number d is the ratio of the geometric progression.

We note that both **the arithmetic and the geometric progressions, are first order recursive sequences, where the x_{n+1} term can indeed, be expressed as a function of n**, (Equations (12-4) and (12-6) respectively).

Example 12-1.
Consider the first order recursive sequence $\{ x_{n+1} = (x_n)^2, \ x_1 = c \}$ (c is given), and express x_{n+1} in terms of n.

Solution

$$x_1 = c, \quad x_2 = (x_1)^2 = c^2, \quad x_3 = (x_2)^2 = c^4, \quad x_4 = (x_3)^2 = c^8, \quad x_5 = (x_4)^2 = c^{16}, \cdots$$

and in general, $x_{n+1} = c^{2^n}$.

Example 12-2.
Consider the first order recursive sequence $\{ x_{n+1} = 2 \cdot \frac{x_n+1}{x_n+3}, \ x_1 = c \}$.

The equation $y = 2 \cdot \frac{y+1}{y+3}$ is called **the characteristic equation** of the given sequence. Show that if the first term $x_1 = c$, is a root of the characteristic equation, then the sequence (x_n) is **a constant sequence**, i.e., $c = x_1 = x_2 = x_3 = x_4 = \cdots$

Solution

Assuming that the first term $x_1 = c$ is a root of the characteristic equation of the sequence, i.e. $c = 2 \cdot \frac{c+1}{c+3}$, the second term x_2 will be,

$$x_2 = 2 \cdot \frac{x_1+1}{x_1+3} = 2 \cdot \frac{c+1}{c+3} = c, \text{ the third term}$$

$$x_3 = 2 \cdot \frac{x_2+1}{x_2+3} = 2 \cdot \frac{c+1}{c+2} = c, \text{ and similarly, } x_4 = c, \; x_5 = c, \ldots \; i.e. \; x_n = c, \; \forall n \in \mathbb{N}.$$

Example 12-3.

The characteristic equation of the sequence, in Example 12-2, has two roots, $y_1 = 1$ and $y_2 = -2$. Let us now assume that $x_1 \notin \{1, -2\}$, (otherwise (x_n) would be a constant sequence as shown in the previous Example), and let us take, for example, $x_1 = 3$. Our sequence now is, $\{ x_{n+1} = 2 \cdot \frac{x_n+1}{x_n+3}, \; x_1 = 3 \}$, and the problem is **to express the $n^{\underline{th}}$ term x_n in terms of n.**

Solution

Let us consider the sequence $b_n = \frac{x_n - y_1}{x_n - y_2} = \frac{x_n - 1}{x_n + 2}$. (*)

We note that,

$$b_{n+1} = \frac{x_{n+1}-1}{x_{n+1}+2} = \frac{2 \cdot \frac{x_n+1}{x_n+3} - 1}{2 \cdot \frac{x_n+1}{x_n+3} + 2} = \frac{1}{4} \cdot \frac{x_n - 1}{x_n + 2} = \frac{1}{4} \cdot b_n. \qquad (**)$$

The sequence (b_n) is **a geometric progression**, with ratio $\frac{1}{4}$, (see (12-5)),therefore, from (12-6), we have,

$$b_{n+1} = b_1 \cdot \left(\frac{1}{4}\right)^n = \frac{x_1 - 1}{x_1 + 2} \cdot \left(\frac{1}{4}\right)^n = \frac{3-1}{3+2} \cdot \left(\frac{1}{4}\right)^n = \frac{2}{5} \cdot \left(\frac{1}{4}\right)^n,$$

or equivalently,

$$\frac{x_{n+1}-1}{x_{n+1}+2} = \frac{2}{5}\cdot\left(\frac{1}{4}\right)^n \Leftrightarrow x_{n+1} = \frac{5\cdot 4^n+4}{5\cdot 4^n-2}\cdot \qquad\qquad (***)$$

The general term x_n has thus been expressed in terms of n.

Note: **a)** This approach can be applied to any first order recursive sequence of the form,

$$\{x_{n+1} = \frac{Ax_n+B}{Kx_n+L}\ ,\ x_1 = c\}$$

where A, B, K, L and c are known constants.

b) For cases where **the characteristic equation** $y = \frac{Ay+B}{Ky+L}$, **has a double root,** $(y_1 = y_2)$, see Problem 12-7.

Example 12-4.

Consider the sequence $\{\ x_{n+1} = \frac{1}{2}\cdot\left(x_n + \frac{4}{x_n}\right),\ x_1 = c\ \}$.

The equation $y = \frac{1}{2}\cdot\left(y + \frac{4}{y}\right)$ is known as **the characteristic equation** of the given sequence. Show that if the first term $x_1 = c$, is a root of the characteristic equation, the sequence (x_n) is **a constant sequence,** i.e. $c = x_1 = x_2 = x_3 = \cdots$ (Compare with Example 12-2).

Solution

Assuming that $x_1 = c$, is a root of the characteristic equation, i.e $c = \frac{1}{2}\cdot\left(c + \frac{4}{c}\right)$, we have, $x_2 = \frac{1}{2}\cdot\left(x_1 + \frac{4}{x_1}\right) = \frac{1}{2}\cdot\left(c + \frac{4}{c}\right) = c$, $x_3 = \frac{1}{2}\cdot\left(x_2 + \frac{4}{x_2}\right) = \frac{1}{2}\cdot\left(c + \frac{4}{c}\right) = c$, and similarly, $x_4 = c,\ x_5 = c, \cdots$, i.e. $x_n = c,\ \forall n \in \mathbb{N}$.

Example 12-5.

Consider the sequence $\{x_{n+1} = \frac{1}{2}\cdot\left(x_n + \frac{4}{x_n}\right),\ x_1 = 3\}$ and **express the $n^{\underline{th}}$ term** x_n in terms of n.

Solution

The characteristic equation of the given sequence (x_n), is $y = \frac{1}{2} \cdot \left(y + \frac{4}{y} \right)$, and has two roots, $y_1 = 2$, $y_2 = -2$. We note that the first term $x_1 = 3$, is different from the two roots. Let us again, consider the sequence,

$$b_n = \frac{x_n - y_1}{x_n - y_2} = \frac{x_n - 2}{x_n + 2}.$$

We note that,

$$b_{n+1} = \frac{x_{n+1} - 2}{x_{n+1} + 2} = \frac{\frac{1}{2}\left(x_n + \frac{4}{x_n}\right) - 2}{\frac{1}{2}\left(x_n + \frac{4}{x_n}\right) + 2} = \left(\frac{x_n - 2}{x_n + 2}\right)^2 = (b_n)^2.$$

The sequence (b_n) was studied in Example 12-1, so

$$b_{n+1} = (b_1)^{2^n} = \left(\frac{x_1 - 2}{x_1 + 2}\right)^{2^n} = \left(\frac{3-2}{3+2}\right)^{2^n} = \left(\frac{1}{5}\right)^{2^n},$$

or equivalently,

$$\frac{x_{n+1} - 2}{x_{n+1} + 2} = \left(\frac{1}{5}\right)^{2^n} \Leftrightarrow x_{n+1} = 2 \cdot \frac{5^{2^n} + 1}{5^{2^n} - 1}.$$

Note: Any first order recursive sequence, of the form,

$$\left\{ x_{n+1} = \frac{1}{2} \cdot \left(x_n + \frac{A}{x_n} \right), \quad x_1 = c \right\}$$

where A and c are given constants can be treated similarly.

Example 12-6.

Consider the sequence $x_{n+1} = a \cdot x_n + b$, $x_1 = c$, where a, b, c are constants, $a \notin \{0, 1\}$.

Express the $n^{\underline{th}}$ term x_n in terms of n.

Solution

The cases where $a = 0$, or $a = 1$, are trivial, ($a = 0$ means that $x_1 = c$ and $b = x_2 = x_3 = x_4 = x_5 = \cdots$, while $a = 1$ means that (x_n) is an arithmetic progression).

The characteristic equation, in this case is $y = ay + b \Leftrightarrow y = \frac{b}{1-a}$.

As in the previous Examples, it is easily shown that **if $x_1 = c$, is a root of the characteristic equation, then (x_n) is a constant sequence**. We therefore focus to the case where $x_1 = c \neq \frac{b}{1-a}$.

Subtracting the characteristic equation from the given sequence, term wise, we obtain,

$$x_{n+1} - y = a \cdot (x_n - y) = a \cdot \left(x_n - \frac{b}{1-a} \right),$$

meaning that the sequence $(x_n - y)$ is **a geometric progression**, therefore,

$$x_{n+1} - y = (x_1 - y) \cdot a^n \Leftrightarrow x_{n+1} - \frac{b}{1-a} = \left(c - \frac{b}{1-a} \right) a^n \Leftrightarrow$$

$$x_{n+1} = \frac{b}{1-a} + \left(c - \frac{b}{1-a} \right) \cdot a^n.$$

Example 12-7. (Second order recursive sequences, linear with constant coefficients).

Let us consider the sequence

$$\{x_{n+2} = b_1 \cdot x_{n+1} + b_2 \cdot x_n, \quad x_1 = c_1, \quad x_2 = c_2 \},$$

where b_1, b_2, and c_1, c_2 are given (known) real numbers.

The problem is **to express the $n^{\underline{th}}$ term x_n in terms of n.**

Solution

Let us assume that $x_n = r^n, \ (r \neq 0)$, where r **is to be determined, so that r^n identically satisfies the given recursive sequence**, i.e.

$$r^{n+2} = b_1 \cdot r^{n+1} + b_2 \cdot r^n \Leftrightarrow r^n \cdot (r^2 - b_1 \cdot r - b_2) = 0,$$

or since $r \neq 0$,

$$r^2 - b_1 \cdot r - b_2 = 0, \quad \text{(The characteristic equation).} \qquad (*)$$

Equation (*) is known as the characteristic equation of the second order recursive sequence. Since the **characteristic equation is a quadratic equation**, there are three cases.

Case I: The two roots r_1, r_2, are **real, distinct numbers**, $(r_1 \neq r_2)$. Then,

$$x_n = A \cdot r_1{}^n + B \cdot r_2{}^n, \qquad\qquad (**)$$

where A and B are constants to be determined from the initial conditions,

$$\{x_1 = c_1 = A \cdot r_1 + B \cdot r_2 \ , and \ \ x_2 = c_2 = A \cdot r_1{}^2 + B \cdot r_2{}^2 \ \}.$$

(Two equations in two unknowns, A and B).

Case II: The two roots are **real equal numbers**, $(r_1 = r_2 \equiv r)$. Then,

$$x_n = (A + B \cdot n) \cdot r^n , \qquad\qquad (***)$$

where A and B are constants to be determined from the initial conditions,(as in Case I).

Case III: The two roots are **two complex conjugate numbers**, i.e.

$$r_1 = re^{i\phi}, \ \ r_2 = re^{-i\phi}, \quad \left(r = |r_1| = |r_2|, \ \ i = \sqrt{-1}\right).$$

In this case,

$$x_n = A \cdot r_1{}^n + B \cdot r_2{}^n. \qquad\qquad (****)$$

Making use of the famous **Euler's formulas**,

$$e^{i\phi} = \cos\phi + i\sin\phi \ , \ \ e^{-i\phi} = \cos\phi - i\sin\phi, \ \ i = \sqrt{-1},$$

expression (****) may be recast in an equivalent, more convenient form,

$$x_n = r^n \cdot \{F\cos(n\varphi) + G\sin(n\phi)\}, \qquad\qquad (*****)$$

where as in the previous cases, the constants F and G, are to be determined from the two initial conditions, $x_1 = c_1$ and $x_2 = c_2$, (see Problem 12-14).

Example 12-8. (The Fibonacci sequence).

The sequence, $\{x_{n+2} = x_{n+1} + x_n,\ x_1 = 1,\ x_2 = 1\}$ is known as the **Fibonacci sequence**, (see also Example (1-3)). Express the $n^{\underline{th}}$ term x_n in terms of n.

Solution

The characteristic equation, (equation (*) in Example 12-7), is

$$r^2 - r - 1 = 0 \Leftrightarrow r_1 = \frac{1+\sqrt{5}}{2},\quad r_2 = \frac{1-\sqrt{5}}{2},$$

so the general term x_n will be,

$$x_n = A \cdot r_1{}^n + B \cdot r_2{}^n.$$

From the given initial conditions,

$$\{x_1 = A \cdot r_1 + B \cdot r_2 = 1,\quad x_2 = A \cdot r_1{}^2 + B \cdot r_2{}^2 = 1\} \Longrightarrow \left\{ A = \tfrac{1}{\sqrt{5}},\ B = -\tfrac{1}{\sqrt{5}} \right\},$$

and finally, one obtains,

$$x_n = \frac{(1+\sqrt{5})^n - (1-\sqrt{5})^n}{\sqrt{5}\cdot 2^n},\quad n = 1, 2, 3, \dots.$$

Note: The numbers generated by the Fibonacci sequence are called **Fibonacci numbers**. For a relationship between **the Fibonacci numbers** and the **Golden Ratio** $\varphi = \frac{1+\sqrt{5}}{2}$, see Problem 12-9.

Example 12-9.

Consider the sequence $\{x_{n+2} = 2x_{n+1} - 2x_n,\ x_1 = 1,\ x_2 = 2\}$, and express the term x_n in terms of n.

Solution

The characteristic equation is,

$$r^2 - 2r + 2 = 0 \Longrightarrow r_1 = 1 + i,\ r_2 = 1 - i.$$

(Case III, in Example 12-7). In exponential form, $r_1 = \sqrt{2} \cdot e^{i\frac{\pi}{4}}$, $r_2 = \sqrt{2} \cdot e^{-i\frac{\pi}{4}}$, so

$$x_n = \left(\sqrt{2}\right)^n \cdot \left\{ F \cos\left(\frac{n\pi}{4}\right) + G \sin\left(\frac{n\pi}{4}\right) \right\}.$$

From the initial conditions,

$$x_1 = 1 \Longrightarrow \sqrt{2} \cdot \left\{ F \cos\frac{\pi}{4} + G \sin\frac{\pi}{4} \right\} = 1 \Longrightarrow \sqrt{2} \cdot \left\{ F \frac{\sqrt{2}}{2} + G \frac{\sqrt{2}}{2} \right\} = 1 \Longrightarrow F + G = 1.$$

$$x_2 = 2 \Longrightarrow (\sqrt{2})^2 \cdot \left\{ F \cos\frac{\pi}{2} + G \sin\frac{\pi}{2} \right\} = 2 \Longrightarrow 2G = 2 \Longrightarrow G = 1.$$

From the above two initial conditions one obtains easily that $F = 0$, $G = 1$, and finally,

$$x_n = \left(\sqrt{2}\right)^n \sin\left(\frac{n\pi}{4}\right), \qquad n = 1, 2, 3, \ldots.$$

PROBLEMS

12-1) Show that the terms of the Fibonacci sequence, satisfy
$$x_{n+1}{}^2 - x_n \cdot x_{n+2} = (-1)^n.$$

12-2) Consider the sequence $\{ x_{n+2} = 2x_{n+1} - x_n , x_1 = 1, x_2 = 2 \}$, and express the general term x_n in terms of n.
Hint: Case II in Example 12-7.

12-3) Consider the sequence $\{ x_{n+2} = x_{n+1} + x_n + 1, x_1 = 2, x_2 = 4 \}$, and express the term x_n in terms of n.
Hint: Define a new sequence $b_n = x_n + 1$, and show, $\{b_{n+2} = b_{n+1} + b_n , b_1 = 3, b_2 = 5\}$. This is a Fibonacci type sequence, studied in Example 12-8. Once b_n is determined in terms of n, then $x_n = b_n - 1$.
(Answer: $x_n = \frac{5+2\sqrt{5}}{5}\left(\frac{1+\sqrt{5}}{2}\right)^n + \frac{5-2\sqrt{5}}{5}\left(\frac{1-\sqrt{5}}{2}\right)^n - 1$).

12-4) Consider the sequence $\{ x_{n+1} = \frac{x_n+2}{x_n}, \quad x_1 = 3 \}$, and express x_n in terms of n.
Hint: See Example 12-3.

12-5) Consider the sequence $\{ x_{n+1} = \frac{1}{2} \cdot \left(x_n + \frac{1}{x_n} \right), x_1 = 3 \}$. Following the approach developed in Example 12-5, show that $x_{n+1} = \frac{2^{2^n}+1}{2^{2^n}-1}$.

12-6) Given that $\{ 4x_{n+1} = 2x_n + 3, x_1 = 1 \}$, express x_n in terms of n.
Hint: See Example 12-6.

12-7) If $x_{n+1} = \frac{2x_n-1}{x_n}$, $x_1 = 3$, express x_n in terms of n.
Hint: In this problem **the characteristic equation has a double root**, so we cannot apply directly the method developed in Example 12-3. Set $x_n = \frac{w_{n+1}}{w_n}$, and show that $w_{n+2} = 2w_{n+1} - w_n$. Then solve according to Example 12-7, Case II.
(Answer: $x_n = \frac{2n+1}{2n-1}$).

12-8) If $\{x_1, x_2, x_3, x_4, ...\}$ are the Fibonacci numbers obtained in Example 12-8, show that $\lim_{n\to\infty} x_n = +\infty$. Next consider the sequence (y_n), defined by

$$\frac{1}{y_{n+2}} = \frac{1}{y_{n+1}} + \frac{1}{y_n}, \quad y_1 = 1, y_2 = 1,$$

and show that $\lim_{n\to\infty} y_n = 0$.

12-9) If $\{x_1, x_2, x_3, x_4, ...\}$ are the **Fibonacci numbers** obtained in Example 12-8, and φ is the **Golden ratio** ($\varphi = \frac{1+\sqrt{5}}{2}$), show that $\lim \frac{x_{n+1}}{x_n} = \varphi$.

12-10) If $\{x_1, x_2, x_3, x_4, ...\}$ are the **Fibonacci numbers** obtained in Example 12-8, show that $\lim \sqrt[n]{x_n} = \varphi$.
Hint: Make use of Theorem 10-10, and Problem 12-9.

12-11) If $\{ x_{n+2} = 8x_{n+1} - 25x_n, x_1 = -2, x_2 = 9 \}$, show that

$$x_n = 5^{n-1} \left(\frac{17}{3} \sin\{(n-1)\varphi\} - 2\cos\{(n-1)\phi\} \right), \text{ where } \phi = Arctan\left(\frac{3}{4}\right).$$

12-12) If $\{ x_{n+1} = \frac{x_n+3}{-x_n+1}, x_1 = 2 \}$, express x_n in terms of n, and show that $x_{n+3} = x_n$, $\forall n$. Which is the term x_{856}? The term x_{1734}?
Hint: In this problem, the characteristic equation has complex roots.

12-13) (Third order recursive sequence).
Consider the sequence $\{ x_{n+3} = 2x_{n+2} + x_{n+1} - 2x_n, \ x_1 = 1, \ x_2 = 1, \ x_3 = -1 \}$, and express x_n in terms of n.

Hint: Following the approach developed in Example 12-7, assume **a trial solution of the form** $x_n = r^n$, and determine r so that r^n identically satisfies the given recursive sequence. **The characteristic equation will be a third degree polynomial in** r, **having roots,** r_1, r_2, r_3. Then, $x_n = Ar_1{}^n + Br_2{}^n + Cr_3{}^n$, where A, B, C are to be determined from the given initial values x_1, x_2, and x_3.

(Answer: $x_n = 2 + \frac{(-1)^n - 2^n}{3}$).

12-14) In Example 12-7, (provided that r_1, r_2 are the two roots of the characteristic equation), show that,

a) In Case I, x_n in equation (**) satisfies the recursive sequence for any values of the constants A and B,

b) In Case II, x_n in equation (***) satisfies the recursive sequence for any values of the constants A and B,

c) Making use of Euler's formulas, expressions (****) and (*****), are completely equivalent, and that x_n in equation (*****) satisfies the recursive sequence for any values of the constants F and G.

12-15) Consider the sequence,

$$\left\{ x_1 = \sqrt{c}, \ x_2 = \sqrt{c + \sqrt{c}}, \ x_3 = \sqrt{c + \sqrt{c + \sqrt{c}}}, \dots, \qquad c > 0 \right\}.$$

Show that the sequence x_n is increasing and bounded, and that $\lim x_n = \frac{1 + \sqrt{1 + 4c}}{2}$.

Hint: See Example 2-5 and Problem 3-6.

12-16) If $\varphi = \frac{1 + \sqrt{5}}{2}$, is the Golden Ratio, show that

$$\varphi = \sqrt{1 + \sqrt{1 + \sqrt{1 + \sqrt{1 + \cdots}}}}$$

Hint: Apply Problem 12-15, with $c = 1$.

12-17) Consider the sequence (x_n) in Example 12-3.

a) Show that (x_n) is decreasing and bounded, and find its limit.

b) Find the $\lim x_n$, from (***), in Example 12-3 and compare your answers.
(Answer: 1).

12-18) Does the sequence (x_n), in Example 12-9, approaches a finite limit, as $n \to \infty$?

12-19) Consider the sequences (x_n) and (y_n), defined by

$$\{x_{n+1} = x_n + 2y_n(\sin \theta)^2 , \quad y_{n+1} = y_n + 2x_n(\cos \theta)^2 , \quad x_1 = 0, \quad y_1 = \cos \theta\}.$$

Show that,

$$x_n = \frac{1}{2}\sin \theta \, \{(1 + \sin 2\theta)^{n-1} - (1 - \sin 2\theta)^{n-1}\},$$

$$y_n = \frac{1}{2}\cos \theta \, \{(1 + \sin 2\theta)^{n-1} + (1 - \sin 2\theta)^{n-1}\}.$$

Hint: Solve the first one for y_n, substitute in the second, etc.

12-20) Given that

$$\{x_{n+1} = 5x_n + y_n , \quad y_{n+1} = 2x_n + 3y_n, \quad x_1 = 1, \quad y_1 = 1\}$$

express x_n and y_n in terms of n.

12-21) Consider the sequences

$$\left\{x_{n+1} = \frac{x_n + y_n}{2} , \quad y_{n+1} = \frac{2x_n y_n}{x_n + y_n}, \quad x_1 > y_1 > 0 \right\}.$$

Express x_n and y_n in terms of n.

Hint: The product $x_{n+1}y_{n+1} = x_n y_n = \cdots = x_1 y_1$. **Set** $c = x_1 y_1$ **(known).**

Then $y_n = \frac{c}{x_n}$, and $x_{n+1} = \frac{1}{2}\left(x_n + \frac{c}{x_n}\right)$, (see Example 12-5), etc.

12-22) Consider the sequence $\{x_{n+2} = 2x_{n+1} + x_n + 1, \quad x_1 = 1, \quad x_2 = 2\}$ and express x_n in terms of n.

Hint: Set $b_n = x_n + \frac{1}{2}$ and show that $b_{n+2} - 2b_{n+1} - b_n = 0$, from which b_n can be expressed in terms of n, and then $x_n = b_n - \frac{1}{2}$.

12-23) Consider the sequence $\left\{ x_{n+1} = \frac{x_n^2 + 2}{2x_n + 1}, \quad x_1 > 1 \right\}$ and show that $x_n > 1 \quad \forall n$, (x_n) is decreasing, and that $\lim x_n = 1$.

12-24) Consider the sequence $\left\{ x_{n+1} = \sqrt[3]{1 + 2x_n} - 1, \quad x_1 = \frac{7}{2} \right\}$ and show that $x_n > 0 \quad \forall n$, (x_n) is decreasing, and that $\lim x_n = 0$.

12-25) Show that the sequence $\left\{ x_{n+1} = \frac{3x_n}{1 + 2x_n}, \quad x_1 = \frac{3}{4} \right\}$ is increasing, bounded and that $\lim x_n = 1$.

13. Cauchy Sequences.

Definition 13-1: A sequence (x_n) is called **a Cauchy sequence**, if for **every positive number $\varepsilon > 0$, there is a positive integer $N = N(\varepsilon)$**, such that

$$\forall n > N \ \text{and} \ \forall m > N \implies |x_n - x_m| < \varepsilon. \qquad (13\text{-}1)$$

A Cauchy sequence is sometimes called **a fundamental sequence.**
Definition 13-1 can be stated in a slightly different, but equivalent form, as follows:

Definition 13-2: A sequence (x_n) is called **a Cauchy sequence**, if for **every positive number $\varepsilon > 0$, there is a positive integer $N = N(\varepsilon)$**, such that

$$\forall n > N \ \text{and} \ \forall k \geq 1 \ (k = 1,2,3,\dots) \implies |x_{n+k} - x_n| < \varepsilon. \ (13\text{-}2)$$

As an example of a Cauchy sequence, let us consider the sequence $x_n = \frac{1}{n}$, $n = 1,2,3,\dots$ Then $|x_m - x_n| = \left|\frac{1}{m} - \frac{1}{n}\right| < \frac{1}{m} + \frac{1}{n}.$ $\qquad (*)$

Let now $\varepsilon > 0$ **be any given positive number**. The inequality $|x_m - x_n| < \varepsilon$ will be satisfied, if $\frac{1}{m} < \frac{\varepsilon}{2}$ and $\frac{1}{n} < \frac{\varepsilon}{2}$, or equivalently, $m > \frac{2}{\varepsilon}$ and $n > \frac{2}{\varepsilon}.$

If we choose $N = \left[\frac{2}{\varepsilon}\right] + 1$, (for the symbol $[x]$, see Example 5-1), then $\forall m > N$ **and** $\forall n > N$, the inequality (*) is satisfied, therefore the sequence (x_n) is a Cauchy sequence.

The following Theorem, which we state without proof, is of fundamental importance, especially in the theoretical investigation of sequences, since **it enables us to prove that a given sequence is convergent without knowing its limit**. Once convergence is established, the limit of the sequence is determined with the aid of Theorems and techniques developed in previous Chapters.

Theorem 13-1.
A sequence (x_n) converges if and only if it is a Cauchy sequence.

We note that this Theorem provides **a necessary and sufficient criterion**, for the convergence of a given sequence (x_n).

The main Theorem 13-1 has some important implications, stated below, in the form of Theorems.

Theorem 13-2.

Let (x_n) be a given sequence such that
$|x_{n+1} - x_n| \leq Aq^n$ where $A > 0$ is a positive constant, and $0 < q < 1$.
Then the sequence (x_n) is convergent.

Proof: It suffices to show that the terms of (x_n) satisfy (13-2). We have,

$$
\begin{aligned}
|x_{n+k} - x_n| &= |(x_{n+k} - x_{n+k-1}) + (x_{n+k-1} - x_{n+k-2}) + \cdots (x_{n+1} - x_n)| \\
&\leq |x_{n+k} - x_{n+k-1}| + |x_{n+k-1} - x_{n+k-2}| + \cdots + |x_{n+1} - x_n| \\
&\leq Aq^{n+k-1} + Aq^{n+k-2} + \cdots + Aq^n = Aq^n\{q^{k-1} + q^{k-2} + \cdots + 1\} \\
&= Aq^n \frac{1 - q^k}{1 - q} < \frac{A}{1 - q} q^n.
\end{aligned}
$$

Since by assumption $0 < q < 1$, $\lim q^n = 0$, (see Theorem 8-1), and therefore $\lim \left(\frac{A}{1-q} q^n \right) = 0$. If $\varepsilon > 0$ is any positive number, arbitrarily small, then $\forall n > N(\varepsilon)$ $\left| \frac{A}{1-q} q^n \right| = \frac{A}{1-q} q^n < \varepsilon$, and consequently, for all $k \geq 1$, $|x_{n+k} - x_n| < \varepsilon$, so (x_n) is a **Cauchy sequence**, and hence converges by Theorem 13-1.

Note: In Example 13-5, we show that $q^{k-1} + q^{k-2} + \cdots q + 1 = \frac{1-q^k}{1-q}$.

Theorem 13-3.

Let (x_n) be a given sequence. If there is a constant number $q, (0 < q < 1)$, such that
$|x_{n+2} - x_{n+1}| \leq q \cdot |x_{n+1} - x_n|,$ $\hspace{2cm}$ (13-3)
then the sequence (x_n) converges.

Proof: $|x_{n+1} - x_n| \leq q|x_n - x_{n-1}| \leq q^2|x_{n-1} - x_{n-2}| \leq \cdots \leq q^{n-1}|x_2 - x_1| = \frac{|x_2 - x_1|}{q} q^n$, and application of Theorem 13-2, with $A = \frac{|x_2 - x_1|}{q}$ shows that (x_n) is convergent.

Theorems 13-2 and 13-3 are used quite often in practice, in order to show that a given sequence converges.

Example 13-1.

Show that the sequence $x_n = 1 - \frac{1}{1!} + \frac{1}{2!} - \frac{1}{3!} + \cdots + \frac{(-1)^n}{n!}$ converges.

Solution

We consider the difference,

$$|x_{n+1} - x_n| = \left|\frac{(-1)^{n+1}}{(n+1)!}\right| = \frac{1}{(n+1)!} < \frac{1}{2^n} \qquad \forall n \geq 2.$$

(For a proof see Problem 13-2).

Application of Theorem 13-2, with $A = 1$, $q = \frac{1}{2}$ shows that (x_n) is convergent.
The limit of this sequence is the number e^{-1},(see (11-7)).

Example 13-2.

Consider the sequence $\{ x_{n+1} = \frac{1}{1+x_n}, \quad x_1 = 1 \}$. Show that $\frac{1}{2} < x_n < 1 \quad \forall n > 2$, and that (x_n) converges. What is the $\lim x_n$?

Solution

Using mathematical induction, one may easily prove that $\frac{1}{2} < x_n < 1 \quad \forall n > 2$, (let the reader prove it). We note that,

$$x_{n+2} - x_{n+1} = \frac{1}{1+x_{n+1}} - \frac{1}{1+x_n} = \frac{-(x_{n+1}-x_n)}{(1+x_{n+1})(1+x_n)},$$

from which one easily obtains that

$$|x_{n+1} - x_n| = \frac{1}{(1+x_{n+1})(1+x_n)}|x_{n+1} - x_n| . \qquad\qquad (*)$$

Since for all $n > 2$, $\frac{1}{2} < x_n < 1$, $\frac{3}{2} < 1 + x_n < 2$, $\frac{3}{2} < 1 + x_{n+1} < 2$ and hence

$$\frac{1}{4} < \frac{1}{(1+x_{n+1})\cdot(1+x_n)} < \frac{4}{9}, \qquad \forall n > 2. \qquad\qquad (**)$$

From (*) and (**) one obtains,

$$|x_{n+2} - x_{n+1}| < \frac{4}{9} \cdot |x_{n+1} - x_n| \qquad \forall n > 2, \quad \left(q = \frac{4}{9} < 1\right),$$

and by Theorem 13-3, (x_n) converges. Let $\lim x_n = \ell$. Taking the limit of both sides of the recursive sequence, and noting that $\lim x_n = \lim x_{n+1} = \ell > 0$, we have,

$$\lim x_{n+1} = \lim \frac{1}{1+x_n} = \frac{1}{1+\lim x_n} \Rightarrow \ell = \frac{1}{1+\ell} \Leftrightarrow \ell^2 + \ell - 1 = 0,$$

from which the positive root is , $\ell = \frac{\sqrt{5}-1}{2}$. **(The negative root is rejected, since $\ell > 0$).**

Example 13-3.

Show that the sequence (x_n) defined by $\{x_{n+2} = \frac{1}{2} \cdot (x_{n+1} + x_n),\ x_1 = 0,\ x_2 = 1\}$ converges. To find the $\lim x_n$ see Problem 13-5.

Solution

If we consider the difference,

$$|x_{n+2} - x_{n+1}| = \left|\frac{1}{2} \cdot (x_{n+1} + x_n) - x_{n+1}\right| = \frac{1}{2} \cdot |x_{n+1} - x_n| < \frac{1}{1.5} \cdot |x_{n+1} - x_n|,$$

$(q = \frac{1}{1.5} < 1)$, and by Theorem 13-3 the sequence (x_n) converges.

Example 13-4.

If n is any positive integer show that $n^2 < 3^n$, and then apply Theorem 13-2, to show that the sequence $x_n = \frac{1}{3} - \frac{2}{3^2} + \frac{3}{3^3} + \cdots + (-1)^{n-1} \frac{n}{3^n}$ is convergent. What is the limit of (x_n)?

Solution

We shall use **Mathematical Induction** to show that $n^2 < 3^n$. For $n = 1$ and $n = 2$ the inequality is obviously true. **Assuming that it is true for $n = k \geq 2$, we shall prove that it will be true for $n = k + 1$ as well**. Starting with our **assumption that $k^2 < 3^k$** we have

$$3^{k+1} > 3k^2 \Rightarrow 3^{k+1} > (k+1)^2 + (k-1)^2 + (k^2 - 2) > (k+1)^2,$$

$\forall k \geq 2$,

and according to the principle of Mathematical Induction, the proposition is true for all positive integers.

Let us now consider the difference

$$|x_{n+1} - x_n| = \left|(-1)^n \frac{n+1}{3^{n+1}}\right| = \frac{n+1}{3^{n+1}} < \frac{1}{(\sqrt{3})^{n+1}} = \frac{1}{\sqrt{3}} \cdot \left(\frac{1}{\sqrt{3}}\right)^n,$$

which is shown easily if the inequality just proved is taken into consideration, and application of Theorem 13-2, with $A = \frac{1}{\sqrt{3}}$ and $q = \frac{1}{\sqrt{3}} < 1$, shows that (x_n) is convergent. To find the limit of (x_n) we proceed as follows:

We start with $x_n = \frac{1}{3} - \frac{2}{3^2} + \frac{3}{3^3} + \cdots + (-1)^{n-1} \frac{n}{3^n}$,   (*)

and multiplying both sides by $\left(-\frac{1}{3}\right)$ yields,

$$-\frac{1}{3} x_n = -\frac{1}{3^2} + \frac{2}{3^3} + \frac{3}{3^4} - \cdots + (-1)^n \frac{n}{3^{n+1}}.$$ (**)

Subtracting (**) from (*) implies,

$$\frac{4}{3} x_n = \frac{1}{3} - \frac{1}{3^2} + \frac{1}{3^3} - \frac{1}{3^4} + \cdots + (-1)^n \frac{1}{3^{n+1}}$$

$$= \frac{1}{3}\left(1 - \frac{1}{3} + \frac{1}{3^2} - \frac{1}{3^3} + \cdots + (-1)^n \frac{1}{3^n}\right) = \frac{1}{3} \cdot \frac{1 - \left(\frac{1}{3}\right)^n}{1 - \left(-\frac{1}{3}\right)}$$

$$= \frac{1}{4} \cdot \left(1 - \left(\frac{1}{3}\right)^n\right) \Rightarrow x_n = \frac{3}{16} \cdot \left(1 - \left(\frac{1}{3}\right)^n\right),$$

(see Example 13-5). As $n \to \infty$, $\lim \left(-\frac{1}{3}\right)^n = 0$, by Theorem 8-1, and hence $\lim x_n = \frac{3}{16}$.

Example 13-5.

Show that $\sum_{n=0}^{k-1} q^n = 1 + q + q^2 + \cdots + q^{k-2} + q^{k-1} = \frac{1-q^k}{1-q}$.

If $|q| < 1$ show that $\sum_{n=0}^{\infty} q^n = 1 + q + q^2 + q^3 + \cdots + q^n + \cdots = \frac{1}{1-q}$.

Solution

Let $S = 1 + q + q^2 + \cdots + q^{k-2} + q^{k-1}$, where $q \neq 1$.

Then $qS = q + q^2 + q^3 + \cdots + q^{k-1} + q^k$, and subtracting from the first, we get,

$(1-q)S = 1 - q^k$, or equivalently, $S = \frac{1-q^k}{1-q}$. As $k \to \infty$, the $\lim q^k = 0$ by

Theorem 8-1, and hence $\sum_{n=0}^{\infty} q^n = 1 + q + q^2 + q^3 + \cdots + q^n + \cdots = \frac{1}{1-q}$.

Example 13-6.

Apply definition 13-2, to show that the sequence $x_n = \frac{n+4}{2n+3}$ converges.

Solution

Let us consider the difference $x_{n+k} - x_n = \frac{n+k+4}{2(n+k)+3} - \frac{n+4}{2n+3} = \frac{-5k}{(2n+3)(2n+2k+3)}$

where **k is any positive integer ≥ 1**, and taking the absolute values of both sides,

we have, $\quad |x_{n+k} - x_n| = \frac{|-5k|}{|(2n+3)(2n+2k+3)|} = \frac{5k}{(2n+3)(2n+2k+3)} = \frac{5}{2} \cdot$

$\frac{2k}{(2n+3)(2n+2k+3)} =$

$\frac{5}{2} \cdot \frac{(2n+2k+3)-(2n+3)}{(2n+3)(2n+2k+3)} = \frac{5}{2} \cdot \left(\frac{1}{2n+3} - \frac{1}{2n+2k+3} \right) < \frac{5}{2} \cdot \frac{1}{2n+3} . \qquad (*)$

We note that inequality (*) is valid for any integer $k \geq 1$.

If ε is any positive number, arbitrarily small, then $\frac{5}{2} \cdot \frac{1}{2n+3} < \varepsilon$ is satisfied

provided that $> \frac{\frac{5}{2\varepsilon}-3}{2}$, so if we choose $N = \left[\frac{\frac{5}{2\varepsilon}-3}{2} \right] + 1$, the **inequality** $|x_{n+k} -$

$x_n| < \varepsilon$ **will be satisfied for all $n \geq N$ and $\forall k \geq 1$**, and hence $x_n = \frac{n+4}{2n+3}$ is a

Cauchy sequence and therefore by Theorem 13-1 converges.

PROBLEMS

13-1) Making use of Theorem 13-3 show that the sequence

$x_n = 1 + \frac{1}{1!} + \frac{1}{2!} + \cdots + \frac{1}{n!}$ is convergent. (The limit of this sequence is the

Euler's number).

13-2) Show that $\dfrac{1}{(n+1)!} < \dfrac{1}{2^n}$ $\quad \forall n \geq 2$.

13-3) Which of the following are **Cauchy sequences** ?

a) $x_n = (n+1)^2$, $\qquad$ **b)** $y_n = \dfrac{\sin(n+1)}{n}$, $\qquad$ **c)** $w_n = \left(1+\dfrac{1}{n}\right)^{3n}$,

d) $a_n = \dfrac{3n^3+7}{2n^2+5n-3}$, $\qquad$ **e)** $b_n = \sqrt[n]{n} + \left(1-\dfrac{1}{3n}\right)^n$, $\quad$ **f)** $c_n = \dfrac{n^n}{n!}$.

(**Answer:** (b), (c), (e)).

13-4) If $y_n = 1 + \dfrac{x}{1!} + \dfrac{x^2}{2!} + \dfrac{x^3}{3!} + \cdots + \dfrac{x^n}{n!}$ show that (y_n) converges, (x is any real number)

13-5) Consider the sequence (x_n), in Example 13-3, which is a second order recursive sequence. Express the $n^{\underline{th}}$ term x_n in terms of n, and find the $\lim x_n$.

(**Answer:** $x_n = \dfrac{2}{3} + \dfrac{4}{3}\cdot\left(-\dfrac{1}{2}\right)^n$, $\lim x_n = \dfrac{2}{3}$).

13-6) Consider the sequence (x_n) in Example 13-2. Using the approach developed in Example 12-3, express the $n^{\underline{th}}$ term x_n in terms of n, and find the $\lim x_n$.

13-7) Show that the sequence $y_n = \dfrac{1}{5} + \dfrac{2}{5^2} + \dfrac{3}{5^3} + \cdots + \dfrac{n}{5^n}$ is a Cauchy sequence and show that $\lim y_n = \dfrac{5}{16}$.

Note: Work as in Example 13-4.

13-8) Given the infinite decimal $0.a_1 a_2 a_3 a_4 \cdots \equiv 0 + \dfrac{a_1}{10} + \dfrac{a_2}{10^2} + \dfrac{a_3}{10^3} + \dfrac{a_4}{10^4} +$ $\cdots$, where $0 \leq a_i \leq 9$, define the sequence $x_n = 0.a_1 a_2 a_3 a_4 \cdots a_n$, and show that (x_n) converges.

Hint: Apply Theorem 13-2 and note that $0 \leq a_n \leq 9$, $\forall n \in \mathbb{N}$.

13-9) If (x_n) is a Cauchy sequence, show that (x_n^3) is also a Cauchy sequence.

13-10) Is the sequence $y_n = \sqrt[n]{n!}$ a Cauchy sequence?

Hint: Consider the sequence $x_n = n!$ and apply Theorem 10-10.

13-11) Apply Theorem 13-2, to show that $y_n = \dfrac{1}{2^n}$ is a Cauchy sequence.

13-12) Consider the sequence $\left\{ x_{n+1} = \frac{x_n}{2} + \frac{1}{x_n}, \quad x_1 = 1 \right\}$, and show that

$x_{n+1} \geq \sqrt{2}, \forall n \geq 1$. Then apply Theorem 13-3 to show that (x_n) is convergent, and that $\lim x_n = \sqrt{2}$.

13-14) Consider the sequence $x_n = \frac{1}{1^2} + \frac{1}{2^2} + \cdots + \frac{1}{n^2}$ and make use of (13-2) to show that the sequence (x_n) is convergent.(For the sum of this sequence, see Problem 10-16).

Hint: $\frac{1}{n^2} < \frac{1}{n-1} - \frac{1}{n}, \quad \forall n \geq 2.$

14. Accumulation Points, Limit Superior and Limit Inferior.

Let $(x_n) = \{x_1, x_2, x_3, \cdots, x_n, x_{n+1}, \cdots\}$ be a given sequence. The set S consisting of all the terms of the sequence,

$$S = \{x_1, x_2, x_3, \cdots, x_n, x_{n+1}, \cdots\}, \qquad (14\text{-}1)$$

is **an infinite set, i.e. S contains infinitely many elements.**

The set S will be called **bounded**, if there exists **a positive constant number K**, such that for every element $x_n \in S$, the inequality $|x_n| \leq K$ holds, $(n = 1,2,3,\cdots)$. In this case we say that **the corresponding sequence (x_n) is bounded**,(see also Chapter 2).

For example, the set $S_1 = \{\frac{1}{1}, \frac{1}{2}, \frac{1}{3}, \cdots, \frac{1}{n}, \frac{1}{n+1}, \cdots\}$ is **bounded**, since $|x_n| \leq 1, \forall n \in \mathbb{N}$, while the set $S_2 = \{1^2, 2^2, 3^2, \cdots n^2, (n+1)^2, \cdots\}$ is **unbounded**.

Prior to stating some fundamental properties about infinite and bounded sets, we need some preliminary definitions.

Definition 14-1: An ε **neighborhood of a point** ℓ, is the set of all points x,such that

$$|x - \ell| < \varepsilon \Leftrightarrow \ell - \varepsilon < x < \ell + \varepsilon, \qquad (14\text{-}2)$$

where ε is any given positive number. (See also Definition 5-2 and Figure 5-1).

A deleted ε neighborhood of a point ℓ, is the set of all points x, such that

$$0 < |x - \ell| < \varepsilon \Leftrightarrow \ell - \varepsilon < x < \ell + \varepsilon, \ x \neq \ell, \qquad (14\text{-}3)$$

i.e. **where the point ℓ itself is excluded.**

Definition 14-2: A point ℓ is called **an accumulation point or a limit point of a sequence** (x_n), if every deleted ε neighborhood of ℓ, contains terms of the sequence, or in other words, if every deleted ε neighborhood of ℓ, $((\ell - \varepsilon, \ell + \varepsilon))$, contains an infinite number of terms of (x_n).

If (x_n) has one accumulation point ℓ, then (x_n) converges to ℓ, (i.e. $\lim x_n = \ell$), but if (x_n) has more than one accumulation points, then (x_n) does not converge to a limit.

The famous **Weierstrass-Bolzano Theorem**, which we will prove shortly, states that **every real, infinite and bounded set, possesses at least one accumulation point.** Therefore, if the infinite set S in (14-1) is also bounded then according to the Weierstrass-Bolzano Theorem, S will possess at least one accumulation point, which is called **the accumulation point of the sequence** (x_n).

If (x_n) has more than one accumulation points, then **the smallest one is called the lower limit or the limit inferior of (x_n), $(\lim \inf x_n)$, while the greatest one is called the upper limit or the limit superior of (x_n), $(\lim \sup x_n)$.**

A sequence (x_n) is convergent to **a finite number ℓ** if and only if,

$$\lim \sup x_n = \lim \inf x_n = \ell.$$

Definition 14-3: If an infinite number of terms of (x_n) exceed any given (arbitrary) positive number $M > 0$, we say that **$\lim \sup x_n = +\infty$.**

If an infinite number terms of (x_n) is less than $-M$, where again $M > 0$ is any given (arbitrary) positive number, we say that **$\lim \inf x_n = -\infty$.**

Theorem 14-1. (Weierstrass-Bolzano Theorem).

Every infinite and bounded set S of real numbers has at least one accumulation point.

Proof: Since S is assumed to be bounded, we can safely assume that S is contained within a finite interval $[x, y]$. Let us now divide this interval into two equal subintervals. Then at least one of these, denoted by $S_1 = [x_1, y_1]$ contains infinitely many points, since by assumption S contains an infinite number of points. The length of $(y_1 - x_1) = \frac{1}{2}(y - x)$. So,

$$S \supset S_1 \quad \text{and} \quad y_1 - x_1 = \frac{1}{2}(y - x). \qquad (*)$$

We next divide the interval $[x_1, y_1]$ into two equal subintervals, one of which denoted by $S_2 = [x_2, y_2]$ contains again infinitely many points. The length of the interval $(y_2 - x_2)$ is, $(y_2 - x_2) = \frac{1}{2}(y_1 - x_1) = \frac{1}{2^2}(y - x)$, and we have,

$$S \supset S_1 \supset S_2 \quad \text{and} \quad (y_2 - x_2) = \frac{1}{2^2}(y - x). \qquad (**)$$

Continuing this process, we obtain a set of intervals $S_n = [x_n, y_n]$, each contained in the preceding one,(nested intervals),and such that,

$$S \supset S_1 \supset S_2 \supset \cdots \supset S_n \supset \cdots \quad \text{and} \quad (y_n - x_n) = \frac{1}{2^n}(y - x), \qquad (***)$$

from which the $\lim(y_n - x_n) = \lim \left\{ \frac{1}{2^n}(y - x) \right\} = (y - x) \lim \frac{1}{2^n} = 0.$

By virtue of Theorem 10-6, (Note) there exists one, unique number ℓ, common to all the intervals S_k. Every deleted ε neighborhood of ℓ, contains an infinite number of points, therefore ℓ is an accumulation or limit point of S.

If S is the set in (14-1), ℓ **will be an accumulation point of the sequence** (x_n).

Example 14-1.
Find the lim sup x_n and the lim inf x_n , if $x_n = (-1)^n \frac{n}{n+1}$.

Solution

The first few terms of x_n, are $\{ x_1 = -\frac{1}{2}, x_2 = \frac{2}{3}, x_3 = -\frac{3}{4}, x_4 = \frac{4}{5}, \cdots, \}$. We first note that the terms of the sequence alternate. If we consider the subsequence,

$$y_n = x_{2n} = \frac{2n}{2n+1}, \quad the \ \lim y_n = \lim x_{2n} = 1.$$

This means that in every ε neighborhood of 1, $(1 - \varepsilon, 1 + \varepsilon)$, there are infinitely many terms of (y_n), and therefore infinitely many terms of (x_{2n}), i.e. **the number 1 is an accumulation point of (x_n).**

If we now consider the subsequence,

$$w_n = x_{2n+1} = -\frac{2n+1}{2n+2}, \quad the \ \lim w_n = \lim x_{2n+1} = -1,$$

and reasoning similarly, we find that the number -1 is another accumulation point of (x_n).

Finally, the **lim sup x_n = 1** (the greatest) and the **lim inf x_n = -1**, (the smallest).

Example 14-2.

Find the lim sup x_n and the lim inf x_n, if $x_n = n^{n+n(-1)^n}$.

Solution

The first few terms of (x_n) are, $\{ x_1 = 1, x_2 = 2^4, x_3 = 1, x_4 = 4^8, x_5 = 1, \cdots \}$.

The subsequence,

$$y_n = x_{2n} = (2n)^{4n} \to +\infty, \quad as \ n \to \infty,$$

while the subsequence,

$$w_n = x_{2n+1} = (2n+1)^0 = 1 \implies \lim w_n = \lim x_{2n+1} = 1.$$

Reasoning as in Example 14-1,

The **lim sup x_n = $+\infty$**, and the **lim inf x_n = 1**.

Example 14-3.

Find the lim sup x_n and the lim inf x_n of the sequence $x_n = 2 + (-1)^n + \frac{(-1)^n}{n+1}$.

Solution

We may first consider the subsequence

$$y_n = x_{2n} = 2 + 1 + \frac{1}{2n+1} = 3 + \frac{1}{2n+1} \Rightarrow \lim y_n = \lim x_{2n} = 3,$$

and then consider the other subsequence,

$$w_n = x_{2n+1} = 2 - 1 - \frac{1}{2n+2} = 1 - \frac{1}{2n+2} \Rightarrow \lim w_n = \lim x_{2n+1} = 1.$$

Reasoning as in the previous Examples,

The **lim sup x_n = 3**, and the **lim inf x_n = 1**.

PROBLEMS

14-1) Find the lim sup and the lim inf of the following sequences,

a) $x_n = (-1)^n \cdot n$,   **b)** $y_n = \frac{(-1)^{n+1}}{n+2}$

(Answer: a)$+\infty, -\infty$, b)0, 0).

14-2) Find the lim sup and the lim inf of $x_n = n^{\frac{1+(-1)^n}{3}}$.

14-3) Find the lim sup and the lim inf of $x_n = \left(1 + \frac{(-1)^n}{n}\right)^n$.

(Answer: e, $\frac{1}{e}$).

14-4) Find the lim sup and the lim inf of $y_n = \left(1 + \frac{5(-1)^n}{n}\right)^n$.

14-5) Find the lim sup and the lim inf of the following sequences,

a) $x_n = e^{1+(-1)^n}$,   **b)** $y_n = \left(1 + \dfrac{2(-1)^n}{n+1}\right)^n$.

(Answer: a) e^2 , 1 , b) e^2 , e^{-2} **).**

15. Sequences of Complex Numbers.

In the previous chapters we studied the theory of real sequences (x_n), meaning that the terms of the sequence $\{x_1, x_2, x_3, \cdots, x_n, x_{n+1}, \cdots\}$ are all **real numbers**.

In a very similar way, one may also define **sequences of complex numbers**.

The reader of this chapter is assumed to be familiar with the elementary theory of complex numbers, i.e. definition, algebraic operations with complex numbers, the three fundamental representation of complex numbers , (**Cartesian, Polar and Exponential representations**), **the** De Moivre's Theorem, etc.

We shall use the letter $\mathbb{C}$ to denote the set of complex numbers.

Definition 15-1: A sequence $(z_n) = \{z_1, z_2, z_3, \cdots, z_n, z_{n+1}, \cdots\}$ of complex numbers, **is a rule that assigns to each element in the set $\mathbb{N}$, one and only one element in the set $\mathbb{C}$.**

For example, $z_n = i^n$, where of course $i = \sqrt{-1}$ **is the imaginary unit,**
$w_n = \left(\frac{1+i}{2}\right)^n$, $u_n = \left(\frac{n+2i}{3}\right)^n$, are complex sequences.

Definition 15-2: Let $(z_n) = \{z_1, z_2, z_3, \cdots, z_n, z_{n+1}, \cdots\}$ be a sequence of complex numbers. We say that **the sequence (z_n) converges to a limit z** (complex number in general),or that (z_n) **tends to a limit z**, if

$$\forall \varepsilon > 0, \ \exists N = N(\varepsilon) : \quad \forall n > N(\varepsilon) \Rightarrow |z_n - z| < \varepsilon, \qquad (15\text{-}1)$$

and we write, $\lim z_n = z \quad or \quad z_n \to z.$

The sequence (z_n) is called convergent to the limit z. If $z = 0 + i0 = 0$, (z_n) is called a null sequence.

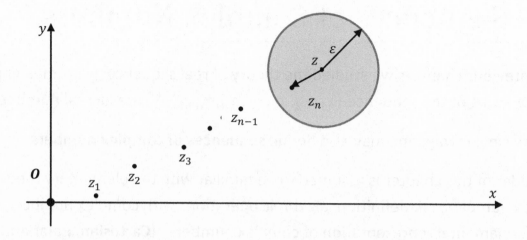

Fig. 15-1: Convergence in the Complex Plane.

Definition 15-2 is very similar to the Definition 6-1 and (15-1) is identical to (6-1),with one exception. **In the complex plane, the ε neighborhood of a point z, i.e. the set of points w,satisfying $0 < |w - z| < \varepsilon$, is a circle having center the point z and radius ε,**(the shaded circle in Fig.15-1).

Therefore, (15-1) simply states that if $\lim z_n = z$, then, having chosen **an arbitrarily small $\varepsilon > 0$**, from a certain stage on, **all the terms of the sequence will lie within the circle centered at z and having radius ε.**

The following theorem is important, since **it reduces the study of complex sequences to the study of real sequences.**

Theorem 15-1.

Let us consider the sequence $z_n = x_n + iy_n$, $x_n \in \mathbb{R}, y_n \in \mathbb{R}$, converging to a complex number $z = x + iy$, $x \in \mathbb{R}$, $y \in \mathbb{R}$. Then,

$$\lim z_n = \lim(x_n + iy_n) = x + iy \Leftrightarrow \{\lim x_n = x, \quad and \quad \lim y_n = y\}. \quad (15\text{-}2)$$

Proof: Let us assume that $\lim z_n = \lim(x_n + iy_n) = x + iy$. This means that,

$$\forall \varepsilon > 0 \; \exists N = N(\varepsilon) : \forall n > N \implies |z_n - z| = |(x_n - x) + i(y_n - y)| < \varepsilon, \quad (*)$$

and since,

$$|x_n - x| \leq |(x_n - x) + i(y_n - y)| = \sqrt{(x_n - x)^2 + (y_n - y)^2},$$

we have, taking (*) into account, that

$$\forall \varepsilon > 0 \quad \exists N = N(\varepsilon) : \forall n > N \implies |x_n - x| < \varepsilon,$$

meaning that $\lim x_n = x$. Reasoning similarly, we can show easily that $\lim y_n = y$.

Conversely, if $\{\lim x_n = x \;\; and \;\; \lim y_n = y\}$, then $\lim z_n = \lim(x_n + iy_n) = x + iy = z$.

For a proof see Problem 15-1.

Theorem 15-2. (Uniqueness of the limit).

The limit of a convergent sequence is unique or stated equivalently,

If $\lim z_n = z_1$, and $\lim z_n = z_2$, then $z_1 = z_2$.

For a proof, see Problem 15-2.

Theorem 15-3.

If $\lim z_n = z$ and $\lim w_n = w$, $((z_n), (w_n)$ are sequences of complex numbers ,and in general z and w are also complex numbers),then,

$$1) \; \lim(c \cdot z_n) = c \cdot \lim z_n = c \cdot z, \quad (c \in \mathbb{C})$$

$$2) \; \lim(z_n + w_n) = \lim z_n + \lim w_n = z + w,$$

$$3) \; \lim(z_n - w_n) = \lim z_n - \lim w_n = z - w,$$

$$4) \; \lim(z_n \cdot w_n) = \lim z_n \cdot \lim w_n = z \cdot w,$$

$$5) \; \lim\left(\frac{z_n}{w_n}\right) = \frac{\lim z_n}{\lim w_n} = \frac{z}{w} \quad (w \neq 0)$$

$$6) \; \lim\left((z_n)^k\right) = (\lim z_n)^k = z^k, (k \in \mathbb{Z}).$$

These properties can obviously be generalized to **any finite number** of sequences, for example,

$$\lim(z_n + w_n + u_n) = \lim z_n + \lim w_n + \lim u_n,$$

$$\lim(z_n \cdot w_n \cdot u_n) = \lim z_n \cdot \lim w_n \cdot \lim u_n,$$

and similarly for the others. For a proof see Problem 15-3.

Example 15-1.

List the first few terms of the sequence, $z_n = i^n$, $(i = \sqrt{-1})$.

Solution

$$\left\{ \begin{array}{l} z_1 = i, z_2 = i^2 = -1, z_3 = i^3 = i^2 i = -i, z_4 = i^4 = i^2 i^2 = (-1)(-1) = 1, z_5 = i^5 = \\ \quad i^4 i = 1i = i, z_6 = i^6 = i^4 i^2 = 1(-1) = -1, \cdots \end{array} \right\}$$

We can easily see that all the elements of $z_n = i^n$ are among the elements of the set $\{i, -i, 1, -1\}$.

Example 15-2.

If $z_n = \frac{1+i}{n}$ find the $\lim z_n$.

Solution

The general term of the sequence is $z_n = \frac{1+i}{n} = \frac{1}{n} + i\frac{1}{n}$ and making use of Theorem 15-1,

$$\lim z_n = \lim\left(\frac{1}{n} + i\frac{1}{n}\right) = \lim\left(\frac{1}{n}\right) + i \lim\left(\frac{1}{n}\right) = 0 + i0 = 0.$$

Therefore (z_n) is **a null sequence**.

Example 15-3.

If $|z| < 1$, show that $z_n = z^n$ is a null sequence.

Solution

It suffices to show that $\forall \varepsilon > 0$, from a certain stage on, (see (15-1)),

$$|z^n - 0| < \varepsilon \Leftrightarrow |z^n| < \varepsilon \Leftrightarrow |z|^n < \varepsilon.$$

However, since the **real number** $|z|$ is less than 1, the last inequality is true,(the sequence $|z|^n$ is null, by Theorem 8-1), and going backwards, $|z^n - 0| < \varepsilon$,meaning that $\lim z^n = 0$.

Example 15-4.
Show that, if $\lim|z_n| = 0$, then $\lim z_n = 0$, and conversely.

Solution

a) If $z_n = x_n + y_n$ then $|z_n| = \sqrt{x_n{}^2 + y_n{}^2}$, therefore,

$\lim|z_n| = 0 \Rightarrow \lim \sqrt{x_n{}^2 + y_n{}^2} = 0 \Rightarrow \lim(x_n{}^2 + y_n{}^2) = 0 \Rightarrow \{\lim x_n = 0, and \lim y_n = 0\}$

and $\lim z_n = \lim(x_n + iy_n) = \lim x_n + i \lim y_n = 0 + i0 = 0$.

b) Conversely, if $\lim z_n = 0 \Rightarrow \lim(x_n + iy_n) = 0 \Rightarrow \lim x_n = 0, and \lim y_n = 0 \Rightarrow \lim(x_n{}^2 + y_n{}^2) = 0 \Rightarrow \lim \sqrt{(x_n{}^2 + y_n{}^2)} \Rightarrow \lim|z_n| = 0$.

Example 15-5.
If $z_n = \dfrac{(\cos\phi + i\sin\phi)^n}{n}$ where ϕ is a given angle in radians, find $\lim z_n$.

Solution

Making use of De Moivre's Theorem,

$z_n = \dfrac{\cos n\phi + i\sin n\phi}{n} \Rightarrow |z_n| = \dfrac{|\cos n\phi + i\sin n\phi|}{n} = \dfrac{\sqrt{(\cos n\phi)^2 + (\sin n\phi)^2}}{n} = \dfrac{1}{n}$.

The $\lim|z_n| = \lim\dfrac{1}{n} = 0$, and as proved in Example 15-4, $\lim z_n = 0$, as well.

Example 15-6.
a) If $z \neq 1$, show that $1 + z + z^2 + z^3 + \cdots + z^n = \dfrac{1 - z^{n+1}}{1-z}$.

b) Assuming further that $|z| < 1$, show that $1 + z + z^2 + \cdots + z^n + \cdots = \dfrac{1}{1-z}$.

Solution

a) If we call $S = 1 + z + z^2 + z^3 + \cdots + z^n$, then   (*)

$$z \cdot S = z + z^2 + z^3 + \cdots + z^n + z^{n+1}.$$ (**)

Subtracting (**) from (*),we have,

$$S \cdot (1 - z) = 1 - z^{n+1} \implies S = \frac{1 - z^{n+1}}{1 - z}.$$

b) Letting $n \to \infty$, $\lim z^{n+1} = 0$, since $|z| < 1$, (see Example 15-3), we have,

$$1 + z + z^2 + z^3 + \cdots + z^n + \cdots = \frac{1}{1 - z}.$$

Example 15-7.

Let (z_n) be a complex sequence, defined **recursively**, by

$$z_{n+1} = \frac{z_n + |z_n|}{2}, \qquad z_1 = \frac{\sqrt{2}}{2} + i\frac{\sqrt{2}}{2}.$$

Express $|z_n|$ as as a function of n.

Solution

If we call $r_n = |z_n| > 0$ and ϕ_n the modulus and the argument of the complex number z_n respectively, then the polar form of z_n will be,

$$z_n = r_n(\cos \phi_n + i \sin \phi_n), \quad n = 1,2,3,\cdots, -\pi \le \phi_n < \pi.$$ (*)

The given recursive formula becomes,

$$r_{n+1}(\cos \phi_{n+1} + i \sin \phi_{n+1}) = \frac{r_n(\cos \phi_n + i \sin \phi_n) + r_n}{2} \quad \text{or,}$$

$$r_{n+1}(\cos \phi_{n+1} + i \sin \phi_{n+1}) = \frac{r_n(1 + \cos \phi_n + i \sin \phi_n)}{2}.$$ (**)

Making use of the well known trigonometric identities,

$$1 + \cos x = 2\left(\cos\frac{x}{2}\right)^2, \qquad \sin x = 2\sin\frac{x}{2}\cos\frac{x}{2},$$ (***)

equation (**) implies,

$$r_{n+1}(\cos\phi_{n+1} + i\sin\phi_{n+1}) = r_n\cos\frac{\phi_n}{2}\left\{\cos\frac{\phi_n}{2} + i\sin\frac{\phi_n}{2}\right\} \Longrightarrow$$

$$\left\{r_{n+1} = r_n\cos\frac{\phi_n}{2}\ \textbf{\textit{and}}\ \phi_{n+1} - \frac{\phi_n}{2} = 2k\pi,\ k = 0, \pm1, \pm2, \pm3, \cdots\right\}.$$
$$(****)$$

However, since $-\pi \le \phi_{n+1} < \pi$, and $-\frac{\pi}{2} < -\frac{\phi_n}{2} \le \frac{\pi}{2}$, the quantity $\phi_{n+1} - \frac{\phi_n}{2}$ will satisfy, $-\frac{3\pi}{2} < \phi_{n+1} - \frac{\phi_n}{2} < \frac{3\pi}{2} \Longleftrightarrow -\frac{3\pi}{2} < 2k\pi < \frac{3\pi}{2} \Longleftrightarrow -\frac{3}{4} < k < \frac{3}{4} \Longrightarrow \boldsymbol{k =}$ **0**, and finally, from the second equation in (****), $\phi_{n+1} = \frac{\phi_n}{2}$.

To summarize, the given recursive formula, finally yields,

$$\left\{r_{n+1} = r_n\cos\frac{\phi_n}{2},\ \phi_{n+1} = \frac{\phi_n}{2}, z_1 = \frac{\sqrt{2}}{2} + i\frac{\sqrt{2}}{2} = \cos\frac{\pi}{4} + i\sin\frac{\pi}{4}\cdot\right\},\quad (*****)$$

therefore,

$$r_2 = r_1\cos\frac{\phi_1}{2},\quad \phi_1 = \frac{\pi}{4}, r_1 = |z_1| = 1$$

$$r_3 = r_2\cos\frac{\phi_2}{2} = r_2\cos\frac{\phi_1}{2^2}$$

$$r_4 = r_1\cos\frac{\phi_3}{2} = r_3\cos\frac{\phi_1}{2^3}$$

$$\cdots\cdots\cdots\cdots\cdots\cdots\cdots\cdots$$

$$r_n = r_{n-1}\cos\frac{\phi_{n-1}}{2} = r_{n-1}\cos\frac{\phi_1}{2^{n-1}}$$

and multiplying term wise, we have,

$$r_n = \cos\frac{\phi_1}{2}\cos\frac{\phi_1}{2^2}\cos\frac{\phi_1}{2^3}\cdots\cos\frac{\phi_1}{2^{n-1}}$$

(since $r_1 = 1$), or using the second equation in (***),

$$r_n = \frac{\sin\phi_1}{2\sin\frac{\phi_1}{2}}\cdot\frac{\sin\frac{\phi_1}{2}}{2\sin\frac{\phi_1}{2^2}}\cdot\frac{\sin\frac{\phi_1}{2^2}}{2\sin\frac{\phi_1}{2^3}}\cdots\frac{\sin\frac{\phi_1}{2^{n-2}}}{2\sin\frac{\phi_1}{2^{n-1}}}$$

which is easily simplified to,

$$r_n = |z_n| = \frac{1}{2^{n-1}} \cdot \frac{\sin\phi_1}{\sin\frac{\phi_1}{2^{n-1}}} = \frac{1}{2^{n-1}} \cdot \frac{\sin\frac{\pi}{4}}{\sin\frac{\pi}{2^{n+1}}} = \frac{\sqrt{2}}{2^n} \cdot \frac{1}{\sin\frac{\pi}{2^{n+1}}} \cdot$$

Example 15-8.

If $\lim z_n = 1 + i$, $\lim w_n = 3 - i$, find $\lim(z_n + w_n)$, $\lim(2iz_n - 7w_n)$, $\lim(z_n w_n)$, $\lim\left(\frac{z_n}{w_n}\right)$, $\lim(z_n^2 \overline{w_n})$, where $\overline{w_n}$ is the complex conjugate of w_n.

Solution

$\lim(z_n + w_n) = \lim z_n + \lim w_n = (1 + i) + (3 - i) = 4,$

$\lim(2iz_n - 7w_n) = (2i)\lim z_n - 7 \lim w_n = 2i(1 + i) - 7(3 - i) = 2i - 2 - 21 + 7i = -23 + 9i,$

$\lim(z_n w_n) = \lim z_n \lim w_n = (1 + i)(3 - i) = 3 + 3i - i + 1 = 4 + 2i,$

$\lim\left(\frac{z_n}{w_n}\right) = \frac{\lim z_n}{\lim w_n} = \frac{1+i}{3-i} = \frac{(1+i)(3+i)}{(3-i)(3+i)} = \frac{1}{5} + i\frac{2}{5},$

$\lim(z_n^2 \overline{w_n}) = \lim z_n^2 \lim \overline{w_n} = (\lim z_n)^2 \overline{\lim w_n} = (1 + i)^2(3 + i) = -2 + 6i.$

PROBLEMS

15-1) If $\lim x_n = x$, and $\lim y_n = y$, show that $\lim(x_n + iy_n) = x + iy$.

15-2) Prove Theorem 1-2.
Hint: Proof similar to the one in Theorem 10-1.

15-3) If $\lim z_n = z$ and $\lim w_n = w$, $((z_n)$ and (w_n) are sequences of complex numbers), show that, $\lim(z_n + w_n) = z + w$, $\lim(z_n - w_n) = z - w$, $\lim(z_n w_n) = z w$, $\lim\left(\frac{z_n}{w_n}\right) = \frac{z}{w}$ $(w \neq 0)$.
Hint: You may use Theorem 15-1.

15-4) In Example 15-7 express z_n in terms of n.

15-5) In Example 15-7, show that $\lim r_n = \lim |z_n| = \frac{2\sqrt{2}}{\pi}$.

Hint: The $\lim_{x \to 0} \frac{\sin x}{x} = 1$.

15-6) Consider the sequence $z_{n+1} = \frac{z_n - |z_n|}{2}$, $z_1 = i$, and express z_n in terms of n.

15-7) If $\lim z_n = i$, $\lim w_n = 1 + i$, $\lim u_n = 3 - i$, find the following limits:

a) $\lim z_n^2$

b) $\lim(z_n w_n u_n)$

c) $\lim \frac{w_n}{u_n}$

d) $\lim(w_n^2 u_n)$

e) $\lim(5z_n + 2w_n - 6u_n)$

f) $\lim \left(\frac{1}{w_n} + u_n \right)$.

(Answer: **a)** -1, **b)** $-2 + 4i$, **c)** $\frac{1+2i}{5}$, **d)** $2 + 6i$, **e)** $-16 + 13i$, **f)** $\frac{7-3i}{2}$).

15-8) Evaluate the infinite sum, $1 + z + z^2 + z^3 + \cdots + z^n + \cdots$, if $z = \frac{1+i\sqrt{2}}{3}$.

Hint: Note that $|z| = \frac{1}{\sqrt{3}} < 1$, and apply formula (b), in Example 15-6.

15-9) Consider the sequence $z_n = \left(\frac{1+i}{\sqrt{2}} \right)^n$, $n = 1,2,3,\cdots$, and show that its terms can take 8 different values only. What is the term z_{1358} ?

Hint: Express z_n in polar form and apply De Moivre's Theorem.

(Answer: $-i$).

15-10) If $\lim z_n = 2 + i$, $\lim w_n = 1 - i$, find the $\lim(\frac{z_n^3}{\bar{w}_n^2})$, where $\bar{w}_n$ is the complex conjugate of w_n.

15-11) Show that $\lim \left(\frac{n}{n+5i} - \frac{in^2}{2n^2+7} \right) = 1 - \frac{i}{2}$.

15-12) Show that $\lim \sqrt{n}\left(\sqrt{n + 5i} - \sqrt{n + 2i} \right) = \frac{3i}{2}$.

15-13) If $\lim z_n = z$, show that $\lim \frac{z_1 + z_2 + z_3 + \cdots + z_n}{n} = z$.

Hint: See proof of Theorem 10-11.

15-14) If $\lim z_n = z$, show that $\lim |z_n| = |z|$. By means of some Examples, show that the converse is not necessarily true.

15-15) If $\left\{ w_{n+1} = \frac{1}{2}\left(w_n + \frac{1}{w_n} \right), \ w_1 = 1 + i \right\}$, express w_{n+1} in terms of n and then show that $\lim w_n = 1$.

Hint: See Example 12-5.

15-16) Making use of Example 15-3, show that the sequences $x_n = \left(\frac{1+i}{5} \right)^n$ and $y_n = \left(\frac{5+2i}{7} \right)^n$ are null sequences.

15-17) Show that the sequence $z_n = \frac{i^n n^3}{2n^3 + 8}$ is bounded, i.e. $|z_n| \leq K$, where K is a constant positive number. Next show that the $\lim z_n$ can take on one of the four different values, $\left\{ \frac{1}{2}, \frac{-1}{2}, \frac{i}{2}, \frac{-i}{2} \right\}$, and from this infer that the $\lim z_n$ does not exist.

15-18) Show that $\lim \dfrac{n^3 + 5n^2 i - 8i}{(in)^3 + 2n + i\sqrt{n}} = i$.

15-19) Show that $\lim \dfrac{\left(2 + \frac{1}{n} \right)^2 - 4}{\left(3 + \frac{i}{n} \right)^2 - 9} = -\dfrac{2i}{3}$.

15-20) In Problem 15-15, if the first term is $w_1 = -1 + i$, show that $\lim w_n = -1$.

16. Special Techniques for Evaluating Limits.

The reader of this Chapter is supposed to be familiar with elements from Differential and Integral Calculus and have a good working knowledge on various related areas.

All Theorems and Techniques developed so far cover a broad area of topics and enable one,

i) To show that a given sequence is convergent or divergent, and

ii) To find the limit of a sequence, (assuming it exists).

However there are sequences, for which the evaluation of the limit, is not an easy task and the so far developed Theorems and Techniques, do not help.

For example, let us consider the sequence $\{x_n = \frac{1}{n+1} + \frac{1}{n+2} + \frac{1}{n+3} + \cdots + \frac{1}{2n}\}$. This sequence is bounded, $(0 < x_n < 1)$, as shown in Example 2-4, and at the same time is increasing as shown in Problem 3-1, therefore by Theorem 10-5, (x_n) converges to a finite limit ≤ 1. However, **since the number of summands of (x_n) depends on n**, the properties developed in Chapter 9, for the evaluation of limits are not applicable,(see **Important Note** in Chapter 9). Other methods of computing the limit of (x_n), and other similar sequences, must be developed.

Some of these methods are outlined below:

A) Evaluation of a limit with the aid of a related Riemann's Integral.

a) Let $x_1, x_2, x_3, \cdots, x_n$ be any n numbers. The **Arithmetic Mean (AM)** or **Average** of these numbers is defined to be, (see also Theorem 10-12),

$$AM = \frac{x_1 + x_2 + x_3 + \cdots x_n}{n}.$$
(16-1)

b) Let us now consider a function $y = f(x)$ defined and continuous over the closed interval $[a, b]$. We divide the interval $[a, b]$ into n subintervals each having width $\Delta x = \frac{b-a}{n}$ and set

$$\{x_0 = a, x_1 = a + \Delta x, x_2 = a + 2\Delta x, \cdots, x_{n-1} = a + (n-1)\Delta x, x_n = b\}$$

Let also,

$$\{y_0 = f(x_0), y_1 = f(x_1), y_2 = f(x_2), \cdots, y_n = f(x_n)\}$$

be the values of the function $y = f(x)$ corresponding to the points $\{x_0, x_1, \cdots, x_{n-1}, x_n\}$. The Arithmetic Mean (**AM**) of the numbers $\{y_1, y_2, \cdots, y_{n-1}, y_n\}$ will be, (according to (16-1)),

$$AM = \frac{y_1 + y_2 + \cdots + y_{n-1} + y_n}{n} = \frac{\sum_{k=1}^{n} y_n}{n} = \frac{\sum_{k=1}^{n} f(a+k\Delta x)}{\left(\frac{b-a}{\Delta x}\right)} = \frac{1}{b-a}\sum_{k=1}^{n} f(a + k\Delta x)\Delta x.$$

If we now consider that $n \to \infty$, or equivalently $(\Delta x) \to 0$, then the last formula above defines **the mean value $\bar{f}$ of the function $y = f(x)$, over the interval $[a, b]$**, i.e.

$$\bar{f} = \frac{1}{b-a} \cdot \lim_{\Delta x \to 0} \sum_{k=1}^{n} f(a + k\Delta x)\Delta x = \frac{1}{b-a} \cdot \int_a^b f(x)\, dx, \quad (16\text{-}2)$$

since by definition, (**Riemann's Definition**),

$$\lim_{\Delta x \to 0} \sum_{k=1}^{n} f(a + k\Delta x)\Delta x = \int_a^b f(x)\, dx.$$

c) If we set $h = b - a$, then $\Delta x = \frac{h}{n}$ and (16-2) may take the equivalent form $\bar{f} = \frac{1}{h}\int_a^{a+h} f(x)dx$, or even more,

$$\lim_{n\to\infty} \left\{\frac{1}{n}\left\{f\left(a + 1\cdot\frac{h}{n}\right) + f\left(a + 2\cdot\frac{h}{n}\right) + \cdots + f\left(a + n\cdot\frac{h}{n}\right)\right\}\right\} = \frac{1}{h}\int_a^{a+h} f(x)dx = \bar{f}.$$

This formula furnishes a convenient way to evaluate limits of the form

$$\lim_{n\to\infty} \left\{\frac{1}{n}\left\{f\left(a + 1\cdot\frac{h}{n}\right) + f\left(a + 2\cdot\frac{h}{n}\right) + \cdots + f\left(a + n\cdot\frac{h}{n}\right)\right\}\right\},$$

in terms of the integral $\frac{1}{h}\int_a^{a+h} f(x)dx$, the evaluation of which is most cases is an easy task, in conjunction with the **Fundamental Theorem of Calculus**,(evaluation of a definite integral in terms of the indefinite integral or the antiderivative of $f(x)$. We remind the reader that **if $F(x)$ is an antiderivative of $f(x)$**, i.e. if $F(x) = \int f(x)dx$, then the definite integral $\int_a^b f(x)\, dx = F(b) - F(a)$, (Newton's-Leibnitz's Theorem).

d) It should also be noted that

$$\lim_{n\to\infty} \left\{\frac{1}{n}\left\{f(a) + f\left(a+1\cdot\frac{h}{n}\right) + \cdots + f\left(a+(n-1)\cdot\frac{h}{n}\right)\right\}\right\} =$$
$$\frac{1}{h}\int_a^{a+h} f(x)dx = \bar{f}.$$

The proof is easy. Let the reader complete the proof.

(For an application of (c) and (d), see Examples 16-1 and 16-2).

B) Evaluation of a limit with the aid of Theorems from Differential Calculus.

e) Suppose now that a sequence $x_n = f(n)$ is given, and that we want to investigate this sequence in regard to its monotonicity. We may consider the **corresponding function $y = f(x)$, and investigate its monotonicity with the aid of derivatives.** We remind that **if within some interval $a \le x < \infty$ the derivative $f'(x) > 0$ then within this interval the function increases, while if $f'(x) < 0$ then within this interval the function $f(x)$ decreases.**
(See Examples 16-4 and 16-5)

g) In order to find the limit of a sequence $a_n = f(n)$ as $n \to \infty$, it suffices to consider the function $y = f(x)$, and find the $\lim_{x\to\infty} f(x)$. **If $\lim_{x\to\infty} f(x) = \ell$, then $\lim_{n\to\infty} a_n = \ell$.**
(See Example 16-3).

Example 16-1.
If $x_n = \frac{1}{n+1} + \frac{1}{n+2} + \frac{1}{n+3} + \cdots + \frac{1}{2n}$ find the $\lim x_n$.

Solution

Let us consider the function $f(x) = \frac{1}{x+1}$, $0 \le x \le 1$. In our case $a = 0, b = 1$ and therefore $h = b - a = 1 - 0 = 1$. If we divide the interval $[0,1]$ into n equal subintervals $\Delta x = \frac{1}{n}$, by means of the points $x_1 = \frac{1}{n}, x_2 = \frac{2}{n}, x_3 = \frac{3}{n}, \cdots, x_{n-1} = \frac{n-1}{n}$, then **the mean value of the function** $f(x) = \frac{1}{x+1}$ **over the interval** $[0, 1]$ will be,

$$\bar{f} = \frac{1}{1}\int_0^1 \frac{1}{x+1} dx = \ln(x+1)|_0^1 = \ln 2 - \ln 1 = \ln 2. \qquad (*)$$

But according to the main formula in part (c) above,

$$\bar{f} = \lim_{n\to\infty}\left\{\frac{1}{n}\left\{\frac{1}{1+\frac{1}{n}} + \frac{1}{1+\frac{2}{n}} + \frac{1}{1+\frac{3}{n}} + \cdots + \frac{1}{1+\frac{n}{n}}\right\}\right\} = \lim_{n\to\infty} x_n. \qquad (**)$$

From (*) and (**), the $\lim_{n\to\infty} x_n = \ln 2$.

Example 16-2.

If $y_n = \frac{n}{n^2+0^2} + \frac{n}{n^2+1^2} + \frac{n}{n^2+2^2} + \cdots + \frac{n}{n^2+(n-1)^2}$, find the $\lim y_n$.

Solution

Let us consider the function $(x) = \frac{1}{1+x^2}$, $0 \le x \le 1$. The mean value of the function $f(x)$ over the interval $[0,1]$ is,

$$\bar{f} = \frac{1}{1}\int_0^1 \frac{1}{1+x^2} dx = \tan^{-1} x|_0^1 = \frac{\pi}{4}. \qquad (*)$$

According to the main formula in part (d) above,

$$\bar{f} = \lim_{n\to\infty}\left\{\frac{1}{n}\left\{\frac{1}{1+\left(\frac{0}{n}\right)^2} + \frac{1}{1+\left(\frac{1}{n}\right)^2} + \cdots + \frac{1}{1+\left(\frac{n-1}{n}\right)^2}\right\}\right\} = \lim_{n\to\infty} y_n. \quad (**)$$

From (*) and (**), the $\lim_{n\to\infty} y_n = \frac{\pi}{4}$.

Example 16-3.

a) If $g(x) = (x + 1)\ln(x + 1) - x$, $x > 0$, show that $g(x) > 0 \;\; \forall x \in \mathbb{R}^+$.

b) Then show that the function $y = f(x) = (x+1)^{1/x}$ is decreasing on the interval $(0, \infty)$ and that $\lim_{x \to \infty} y = 1$.

c) Making use of (a) and (b), show that the sequence $x_n = \sqrt[n]{n+1}$ is decreasing and that $\lim x_n = 1$.

Solution

a) Since $g(0) = 0$, it suffices to show that the derivative $g'(x)$ of $g(x)$ is positive $\forall x \in \mathbb{R}^+$. We have

$g'(x) = (x+1)' \ln(x+1) + (x+1)(\ln(x+1))' - x' = \ln(x+1) + (x+1)\frac{1}{x+1} - 1 = \ln(x+1)$

which is positive $\forall x \in (0, \infty)$.

b) To show that the function $y = (x+1)^{1/x}$ is decreasing on $(0, \infty)$, it suffices to show that within this interval, $y' < 0$. To find the derivative y' we shall use **logarithmic differentiation**. $\ln y = \frac{\ln(x+1)}{x}$ and taking the derivative of both sides, we have,

$$\frac{y'}{y} = \frac{(\ln(x+1))'x - \ln(x+1)x'}{x^2} = \frac{\frac{x}{x+1} - \ln(x+1)}{x^2} = \frac{x - (x+1)\ln(x+1)}{(x+1)x^2},$$

and solving for y' we obtain,

$$y' = y \cdot \frac{x - (x+1)\ln(x+1)}{(x+1)x^2}.$$

Taking part (a) into account, and since $y > 0$ on $(0, \infty)$, the derivative $y' < 0$ over the same interval, meaning that the function $y = (x+1)^{1/x}$ is decreasing on $(0, \infty)$.

To find the $\lim_{x \to \infty} y$ we start with $\ln y = \frac{\ln(x+1)}{x}$ and take the limit of both sides as $x \to \infty$, i.e. $\lim_{x \to \infty} \ln y = \lim_{x \to \infty} \frac{\ln(x+1)}{x} = \frac{\infty}{\infty}$ and making use of the De L'hospital Rule, we have, $\lim_{x \to \infty} \ln y = \lim_{x \to \infty} \frac{(\ln(x+1))'}{x'} = \lim_{x \to \infty} \frac{1}{x+1} = 0$,

and **since the logarithmic function is a continuous function of its argument,**

$\lim_{x\to\infty} \ln y = \ln(\lim_{x\to\infty} y) = 0$, and thus $\lim_{x\to\infty} y = 1$.

c) The terms of the sequence $x_n = \sqrt[n]{n+1}$ are the values of the function $y = (x+1)^{1/x}$ at integers values of the variable x, i.e. $x_n = f(n)$, where $n = 1,2,3,\cdots$. Since $(n+1) > n$, $f(n+1) < f(n)$,(the function $y = f(x)$ is decreasing as shown in (b)), and therefore the sequence $x_{n+1} < x_n$ is decreasing. The $\lim_{n\to\infty} x_n = \lim_{x\to\infty} y = 1$, as proved in (c).

Example 16-4.

Consider the first order recursive sequence $\{x_{n+1} = f(x_n),\ x_1 = c\ \}$, where c is a given constant number. **If the function $y = f(x)$ is increasing in $\mathbb{R}$, show that the sequence (x_n) is monotone,(either increasing or decreasing).**

Solution

Since $y = f(x)$ is increasing in $\mathbb{R}$, by assumption, the derivative $f'(x) > 0$ $\forall x \in \mathbb{R}$.

Let us now consider the difference $x_{n+2} - x_{n+1} = f(x_{n+1}) - f(x_n)$, or making use of the **Mean Value Theorem of Differential Calculus,**

$x_{n+2} - x_{n+1} = f(x_{n+1}) - f(x_n) = (x_{n+1} - x_n)f'(r)$,
where r lies between x_n and x_{n+1}, and since $f'(r) > 0$, it follows that

$\text{sign}(x_{n+2} - x_{n+1}) = \text{sign}(x_{n+1} - x_n) = \cdots = \text{sign}(x_3 - x_2) = \text{sign}(x_2 - x_1)$.

If $\text{sign}(x_2 - x_1) = +$, i.e. if $x_2 > x_1$ then $x_{n+2} > x_{n+1}$ $\forall n$, and **the sequence is increasing**, while if $\text{sign}(x_2 - x_1) = -$, i.e. if $x_2 < x_1$ then $x_{n+2} < x_{n+1}$ $\forall n$, and **the sequence will be decreasing.**

Example 16-5.

Consider the sequence $\{x_{n+1} = 2 + \frac{x_n}{2+x_n},\ x_1 = c > 0\}$. Making use of Example

16-4, find for which values of c the sequence (x_n) is increasing, (decreasing), and then find the $\lim x_n$.

Solution

Let us consider the function $y = 2 + \frac{x}{2+x}$. Since the derivative $f'(x) = \frac{2}{(x+2)^2} > 0, \forall x \in \mathbb{R}$, the function $y = f(x)$ is increasing in $\mathbb{R}$, and therefore according to the Example 16-4, the sequence (x_n) is monotone, i.e.

$$\text{sign}(x_{n+2} - x_{n+1}) = \text{sign}(x_{n+1} - x_n) = \cdots = \text{sign}(x_3 - x_2) = \text{sign}(x_2 - x_1).$$

$$(*)$$

The difference $x_2 - x_1 = 2 + \frac{c}{2+c} - c = \frac{-c^2+c+4}{2+c}$, and

$\text{sign}(x_2 - x_1) = \text{sign}(-c^2 + c + 4)$, since $2 + c > 0$, and therefore,

$\text{sign}(-c^2 + c + 4) = +$, if $0 < c < \frac{1+\sqrt{17}}{2}$, while

$\text{sign}(-c^2 + c + 4) = -$, if $\frac{1+\sqrt{17}}{2} < c < \infty$.

In summary, and taking into consideration (*), **if $0 < c < \frac{1+\sqrt{17}}{2}$ the sequence is increasing, while if $\frac{1+\sqrt{17}}{2} < c < \infty$, the sequence is decreasing**. In both cases the $\lim x_n$ exists and is finite,(for a proof see Problem 16-8). Let $\lim x_n = \ell$. Taking the limits of both sides of the given recursive sequence, and noting that $\lim x_{n+1} = \lim x_n = \ell$, we have,

$$\ell = 2 + \frac{\ell}{2+\ell} \Rightarrow \ell^2 - \ell - 4 = 0 \Rightarrow \ell = \frac{1+\sqrt{17}}{2},$$

and this is the sought for limit, (The other, **negative root** $\frac{1-\sqrt{17}}{2}$ of the quadratic **equation is rejected**, since the limit must be positive).

PROBLEMS

16-1) If $x_n = \frac{\ln(n+1)-\ln n}{\ln(n-1)-\ln n}$, $n = 2,3,4,\cdots$ show that $\lim x_n = -1$.

16-2) If $a_n = n^{\tan(1/n)}$, show that $\lim a_n = 1$.

Hint: Consider the function $y = \left(\frac{1}{x}\right)^{\tan x}$ and find the $\lim_{x\to 0} y$. To find this limit, follow the procedure outlined in Example 16-3.

16-3) If $b_n = \left\{ n \sin\left(\frac{1}{n}\right) \right\}^{n^2}$, show that $\lim b_n = \frac{1}{\sqrt[6]{e}}$.

Hint: Consider the function $y = \left(\frac{\sin x}{x}\right)^{(1/x)^2}$ and find the $\lim_{x\to 0} y$. Follow the procedure outlined in Example 16-3.

16-4) Consider the sequence $x_n = \frac{1^k + 2^k + 3^k + \cdots + n^k}{n^{k+1}}$ where k is a constant positive number, and show that $\lim x_n = \frac{1}{k+1}$.

Hint: Consider the function $f(x) = x^k$, $x \in [0,1]$, and follow the procedure outlined in Examples 16-2 and 16-3.

16-5) Consider the sequence $y_n = \frac{\sqrt[k]{1} + \sqrt[k]{2} + \sqrt[k]{3} + \cdots + \sqrt[k]{n}}{\sqrt[k]{n^{k+1}}}$ where k is any positive integer, and show that $\lim y_n = \frac{k}{k+1}$.

16-6) Show that for quite large values of n, $\sqrt[5]{1} + \sqrt[5]{2} + \sqrt[5]{3} + \cdots + \sqrt[5]{n} \cong \frac{5}{6} n^{(6/5)}$.
Hint: Make use of Problem 16-5, with $k = 5$.

16-7) Show that for quite large values of n, $1^3 + 2^3 + 3^3 + \cdots + n^3 \cong \frac{1}{4} n^4$.

16-8) Show that the sequence in Example 16-5 is bounded, and since it is monotone, as shown in the example, it is convergent.
Hint: It is easy to show that $0 < x_{n+1} < 3$.

16-9) If $c = \frac{1+\sqrt{17}}{2}$ show that (x_n) is a constant sequence, i.e. $x_1 = x_2 = x_3 = \cdots = x_n = \cdots$.

16-10) Show that $\lim_{x\to\infty} \left\{ x \ln \frac{x+1}{x} \right\} = 1$, and then show that $\lim\{n(\ln(n+1) - \ln n)\} = 1$.

16-11) Show that $\lim \left(\cos\frac{1}{n}\right)^{n^2} = \frac{1}{\sqrt{e}}$.

Hint: Consider the function $y = (\cos x)^{\frac{1}{x^2}}$ and find the $\lim_{x \to 0} y$.

16-12) If $b_n = \frac{1}{\sqrt{n^2-0^2}} + \frac{1}{\sqrt{n^2-1^2}} + \frac{1}{\sqrt{n^2-2^2}} + \cdots + \frac{1}{\sqrt{n^2-(n-1)^2}}$ show that $\lim_{n \to \infty} b_n = \frac{\pi}{2}$.

Hint: Consider the average value $\bar{f}$ of the function $f(x) = \frac{1}{\sqrt{1-x^2}}$ on $[0,1]$.

16-13) Consider the sequence $\left\{x_{n+1} = \frac{2+x_n}{3+x_n}, \quad x_1 = c > 0\right\}$. Following the procedure outlined in Example 16-5, find the values of c for which the sequence is increasing (decreasing). What is the $\lim x_n$?

(Answer: Increasing for $0 < c < -1 + \sqrt{3}$, $\lim x_n = -1 + \sqrt{3}$).

16-14) Consider the sequence $x_n = \frac{1!+2!+3!+\cdots+(n-1)!+n!}{(n+1)!}$, and show that (x_n) is decreasing and bounded below (and therefore convergent). Then show that $x_{n+1} = \frac{1}{n+2} + \frac{1}{n+2}x_n$, and finally pass to the limit as $n \to \infty$ to show that $\ell = \lim x_n = 0$.

16-15) If c is any positive number, show that $\lim_{x \to 0} \frac{c^x - 1}{x} = \ln c$. Then consider the sequence $b_n = 2^n\left(\sqrt[2^n]{c} - 1\right)$ and show that $\lim b_n = \ln c$.

16-16) Problem 16-15 suggests a particularly simple method to find the logarithm $\ln c$ of a positive number c. Given c, we take the square root of c n times successively, and then form the number $2^n\left(\sqrt[2^n]{c} - 1\right)$. This number is approximately equal to the $\ln c$. As an example, take $c = 3, n = 8$, find the approximate value of $\ln 3$, and compare with the value of $\ln 3$ obtained using the pocket calculator. Repeat for $c = 5.4$, and $n = 12$.

Made in the USA
Las Vegas, NV
01 May 2021